Lecture Notes in Mathematics

Edited by A. Dold and B. Eckmann

956

Group Actions and Vector Fields

Proceedings of a Polish-North American Seminar
Held at the University of British Columbia
January 15 – February 15, 1981

Edited by J.B. Carrell

Springer-Verlag
Berlin Heidelberg New York 1982

Editor
James B. Carrell
Department of Mathematics, University of British Columbia
Vancouver, B.C. V6T 1Y4, Canada

AMS Subject Classifications (1980): 14D25, 14L30, 14M20, 32M99

ISBN 3-540-11946-9 Springer-Verlag Berlin Heidelberg New York
ISBN 0-387-11946-9 Springer-Verlag New York Heidelberg Berlin

Printing and binding: Beltz Offsetdruck, Hemsbach/Bergstr.
2146/3140-543210

PREFACE

This volume arose from the Polish-North American Seminar on Group Actions and Vector Fields held at the University of British Columbia from January 15 to February 15, 1981. The papers contained herein (with three exceptions) are research papers that were discussed during the seminar. Some of them were not in final form at the time and work was continued during the meeting. The exceptions are the papers of Akyildiz, Dolgachev, and Lieberman. I would like to thank Dolgachev and Lieberman for allowing me to include their papers, both of which are fundamental, but, for some reason or other, have not been published before. The paper of Akyildiz, who was unable to attend, is a generalization of work I reported on.

I would like to express my deepest appreciation to the Natural Sciences and Engineering Research Council of Canada for the support which made this meeting possible. I would also like to thank the Math Department at U.B.C. for its hospitality and for supporting the typing of the manuscripts. Thanks go also to the very pleasant Math Department secretaries, especially Wanda Derksen for their speedy and accurate typing and to Sinan Sertoz for the proofreading.

James B. Carrell

LIST OF PARTICIPANTS

A. Bialynicki - Birula (U. of Warsaw)

J.B. Carrell (U.B.C.)

I. Dolgachev (U. of Michigan)

R. Douglas (U.B.C.)

N. Goldstein (U.B.C., currently Purdue U.)

M. Goresky (U.B.C., currently Northeastern U.)

D. Gross (U. of Notre Dame)

K. Hoechsmann (U.B.C.)

R. Jardine (U.B.C., currently U. of Toronto)

J. Konarski (U. of Warsaw)

M. Koras (U. of Warsaw)

D.I. Lieberman (Inst. Defense Analyses)

L.G. Roberts (U.B.C.)

S. Sertoz (U.B.C.)

A.J. Sommese (U. of Notre Dame)

J. Swiecicka (U. of Warsaw)

P. Wagreich (U. of Ill. at Chicago Circle)

TABLE OF CONTENTS

AKYILDIZ, E. Vector fields and cohomology of G/P 1

BIALYNICKI-BIRULA, A. and SWIECICKA, J. Complete quotients by algebraic torus actions 10

CARRELL, J.B. and SOMMESE, A.J. A generalization of a theorem of Horrocks 23

CARRELL, J.B. and SOMMESE, A.J. Almost homogeneous C^* actions on compact complex surfaces 29

DOLGACHEV, I. Weighted projective varieties 34

KONARSKI, J. A pathological example of an action of k^* 72

KONARSKI, J. Properties of projective orbits of actions of affine algebraic groups 79

KORAS, M. Linearization of reductive group actions 92

LIEBERMAN, D.I. Holomorphic vector fields and rationality 99

SOMMESE, A.J. Some examples of C^* actions 118

WAGREICH, P. The growth function of a discrete group 125

BIBLIOGRAPHY

[illegible] M. Atiyah and I. Macdonald, Commutative Algebra, Addison-Wesley, Reading, Massachusetts.

[Ba] H. Bass, The degree of polynomial growth of finitely generated groups, Proc. London Math. Soc. (3) 25 (1972), 603-614.

[Bo] N. Bourbaki, Groupes et algèbres de Lie, Chapters 4, 5 et 6, Paris, Hermann, 1968.

[illegible] A. Borel, Sur la cohomologie des espaces fibrés principaux et des espaces homogènes de groupes de Lie compacts, Annals of Math., 57 (1953), 115-207.

[illegible] R. Bott, An application of the Morse theory to the topology of Lie groups, Bull. Soc. Math. France, 84 (1956), 251-281.

[Ca1] J. Cannon, The growth of the closed surface groups and compact hyperbolic Coxeter groups (preprint).

[Ca2] J. Cannon, The combinatorial structure of cocompact discrete hyperbolic groups (preprint).

[illegible] P. Cartier, Les arrangements d'hyperplans: un chapitre de géométrie combinatoire, Sém. Bourbaki 1980/81, no. 561.

[Ch] C. Chevalley, Invariants of finite groups generated by reflections, Amer. J. Math., 77 (1955), 778-782.

[Co] A.J. Coleman, The Betti numbers of the simple groups, Can. J. Math., 10 (1958), 349-356.

[illegible] H.S.M. Coxeter, The product of generators of a finite group generated by reflections, Duke Math. Jour., 18 (1951), 765-782.

[illegible] H.S.M. Coxeter and W.O. Moser, Generators and relations for discrete groups, New York, Springer, 1965.

[D] A. Durfee, Fifteen characterizations of rational double points and simple critical points.

[G] M. Gromov, Groups of polynomial growth and expanding maps, I.H.E.S., 1980 (preprint).

[illegible] N. Iwahori and H. Matsumoto, On some Bruhat decomposition and the structure of Hecke rings of p-adic Chevalley groups, Publ. Math. Inst. Hautes Et. Sci., 25 (1965), 5-48.

[illegible] J. Milnor, A note on curvature and the fundamental group, J. Diff. Geometry 2 (1968), 1-7.

[illegible] J. Milnor, Growth of finitely generated solvable groups, J. Diff. Geometry 2 (1968), 447-449.

[illegible] P. Orlik and L. Solomon, Combinatorics and topology of complements of hyperplanes, Invent. Math. 56 (1980), 167-189.

[illegible] P. Orlik and L. Solomon, Unitary reflection groups and cohomology, Invent. Math., 59 (1980), 77-94.

VECTOR FIELDS AND COHOMOLOGY OF G/P

by

Ersan Akyildiz

Abstract

We discuss the cohomology rings of homogeneous spaces from the viewpoint of zeros of vector fields.

I. Introduction

If a compact Kaehler manifold X admits a holomorphic vector field V with isolated zeros, then by a theorem due to J.B. Carrell and D. Lieberman [$C-L_2$] the cohomology ring of X can be calculated around the zeros of V. Although holomorphic vector fields with isolated zeros are not abundant, they do exist on a fundamental class of spaces, namely the algebraic homogeneous spaces.

In this note, the cohomology ring of a homogeneous space G/P together with the cohomology maps of $\pi : G/B \to G/P$, and $i : P/B \to G/B$ will be discussed from the viewpoint of zeros of vector fields. In particular a theorem of A. Borel on the cohomology ring of G/P [B] is obtained rather surprisingly as a limiting case. This description of the cohomology rings and the cohomology maps was a key point in computing the Gysin homomorphism of $\pi : G/B \to G/P$ in [$A-C_1$] and [$A-C_2$].

II. Review of Vector Fields and Cohomology

A holomorphic vector field V on a complex manifold X defines, by way of the contraction operator $i(V)$ a complex of sheaves,

$$0 \to \Omega^n \xrightarrow{i(V)} \Omega^{n-1} \to \cdots \to \Omega^1 \to \mathcal{O} \to 0 .$$

If V has only finitely many zeros, then this complex is exact except at $\mathcal{O}$, and in fact provides a locally free resolution of the sheaf $\mathcal{O}_Z = \mathcal{O}/i(V)\Omega^1$, which is, by definition, the structure sheaf of the variety Z of zeros of V. It follows from the general facts on hypercohomology that there are two spectral sequences $\{'E_r^{p,q}\}$, $\{''E_r^{p,q}\}$ abutting to $\mathrm{Ext}^*(X; \mathcal{O}_Z, \Omega^n)$ where $'E_1^{p,q} = H^q(X, \Omega^{n-p})$, and $''E_2^{p,q} = H^p(X, \underline{\mathrm{Ext}}^q_{\mathcal{O}}(\mathcal{O}_Z, \Omega^n))$. The key fact proved in [$C-L_1$] is that if X is compact Kaehler, then the first spectral sequence degenerates at $'E_1$ as long as $Z \neq \phi$. As a consequence of the finiteness of Z and $H^0(X, \mathcal{O}_Z) \cong \mathrm{Ext}^n(X; \mathcal{O}_Z, \Omega^n)$, where $n = \dim X$, we have the Theorem [$C-L_2$]. If X is a compact Kaehler manifold admitting a holomorphic vector field V with $Z = \text{zero}(V)$ finite but nontrivial, then

(i) $H^p(X,\Omega^q) = 0$ if $p \neq q$

(ii) there exists a filtration

$$H^0(X,\mathcal{O}_Z) = F_n(V) \supset F_{n-1}(V) \supset \ldots \supset F_1(V) \supset F_0(V) \ ,$$

where $n = \dim X$, such that $F_i(V)F_j(V) \subset F_{i+j}(V)$ and having the property that as graded rings

$$(1) \quad m_V : gr(H^0(X,\mathcal{O}_Z)) = \bigoplus_p F_p(V)/F_{p-1}(V) \xrightarrow{\simeq} \bigoplus_p H^p(X,\Omega^p) = H^*(X,C) \ .$$

The key to understanding the isomorphism (1) is in knowing how the Chern classes of a holomorphic vector bundle arise. To answer this we need to recall the theory of V-equivariant bundles [A_1], [A_2], [C-L_2]. For our purpose we only need to discuss the line bundles. We say a holomorphic line bundle L on X is V-equivariant if there exists a V-derivation $\tilde{V} : \mathcal{O}(L) \to \mathcal{O}(L)$; i.e. a C-linear map satisfying $\tilde{V}(fs) = V(f)s + f\tilde{V}(s)$ if $f \in \mathcal{O}$, $s \in \mathcal{O}(L)$. Since $V(f) = i(V)df$, $\tilde{V}$ defines a global section of $\mathrm{End}(\mathcal{O}(L) \otimes_{\mathcal{O}} \mathcal{O}_Z) \cong \mathcal{O}_Z$; i.e. $\tilde{V} \in H^0(X,\mathcal{O}_Z)$. It is shown in [$A_1$], [$A_2$], and [C-$L_2$] that

(a) $\tilde{V} \in F_1(V)$ has image the first Chern class $c_1(L)$ of L under the isomorphism (1), and

(b) if $H^1(X,\mathcal{O}_X) = 0$, then any line bundle L on X is V-equivariant.

The relation between cohomology maps and zeros of vector fields can be explained as follows. Let $f : X \to Y$ be a holomorphic map between compact Kaehler manifolds, V and V^* holomorphic vector fields on X and Y with isolated zeros. If $df(V(x)) = V^*(f(x))$ for any x in X, then it follows from the general facts on hypercohomology and the functoriality of the isomorphism $H^0(X,\mathcal{O}_Z) \cong \mathrm{Ext}^n(X; \mathcal{O}_Z,\Omega^n)$ [G-H, p. 707] that the natural map $f^* : H^0(Y,\mathcal{O}_{Z^*}) \to H^0(X,\mathcal{O}_Z)$ preserves the filtrations, i.e. $f^*(F_p(V^*)) \subseteq F_p(V)$ for each p, and also form the following commutative diagram between the cohomology rings.

$$(2) \qquad \begin{array}{ccc} f^* : H^*(Y,C) & \longrightarrow & H^*(X,C) \\ m_{V^*} \uparrow \wr & & \wr \uparrow m_V \\ gr(f^*) : gr(H^0(Y,\mathcal{O}_{Z^*})) & \longrightarrow & gr(H^0(X,\mathcal{O}_Z)) \ , \end{array}$$

where $Z^* = \mathrm{zero}\,(V^*)$, $Z = \mathrm{zero}\,(V)$, $gr(f^*)$ is the natural graded algebra homomorphism associated to $f^* : H^0(Y,\mathcal{O}_{Z^*}) \to H^0(X,\mathcal{O}_Z)$.

If the vector field V on X has only simple isolated zeros, in other words $Z = \mathrm{zero}\,(V)$ is nonsingular, then $H^0(X,\mathcal{O}_Z) \simeq \bigoplus_{p \in Z} C_p$ is precisely the ring of complex valued functions on Z. Thus, algebraically, $H^0(X,\mathcal{O}_Z)$ can be quite simple. The difficulty in analyzing the cohomology ring is in describing the filtration $F_p(V)$. We will now give a vector field V on G/B inducing vector fields on P/B and G/P with simple isolated zeros. Then by using some facts from Invariant theory we will compute the filtrations induced by these vector fields, and thus the cohomology rings and the cohomology maps.

III. Description of $H^0(G/P,\mathcal{O}_Z)$

We will use the following notation: G will be a connected semisimple linear algebraic group over the field of complex numbers, B a fixed Borel subgroup of G, B_u the unipotent radical of B, H a fixed maximal torus contained in B, g the Lie algebra of G, h and b_u the Lie algebras of H and B_u respectively, $\Delta \subset h^*$ the root system of h in g, Δ_+ the set of positive roots, namely the set of roots of h in b_u, $\Sigma \subset \Delta_+$ the set of simple roots, W the Weyl group of G, $\Theta \subseteq \Sigma$ any subset of Σ, W_Θ the subgroup of W generated by the reflections σ_α, $\alpha \in \Theta$, $P = P_\Theta$ the parabolic subgroup of G corresponding to Θ, $X(H)$ the group of characters of H.

We shall denote by the same symbol an element of $X(H)$ and the corresponding element of Δ ($\alpha = d\alpha$, the differential of $\alpha \in X(H)$ at the identity) when this can be done without any ambiguity. For the basic facts about algebraic groups the reader is referred to [H].

Let $\omega \cdot v = \mathrm{Ad}\omega(v)$ denote the tangent action of W on h, $\omega \in W$, $v \in h$. W thus acts effectively on h and on h^*. Thus we get an action of W on $R = \mathrm{Sym}(h^*)$, the symmetric algebra of h^*, in the usual way: $\omega \cdot f(v) = f(\omega^{-1} \cdot v)$ for $f \in R$. Let R^W be the ring of invariants of W. Since the degree of the natural map $\mathrm{Spec}(R) \to \mathrm{Spec}(R^W)$ is equal to $|W|$, the order of W, there exists a dense open set U in h such that $\omega \cdot v \neq v$ for any $\omega \in W$ and $v \in U$. An element v of U is called a regular vector in h. For a regular vector v in h, $\exp(tv)$ gives a one-parameter subgroup of H so that the fixed point scheme of this action on G/B is exactly $(G/B)^H \cong \{\omega : \omega \in W\}$, where H acts on G/B via the left multiplication. Let $V = \frac{d}{dt}(\exp(tv))\Big|_{t=0}$ be the vector field on G/B associated to this one-parameter family. Then $d\pi(V) = V^*$ is a well defined vector field on G/P, where $\pi : G/B \to G/P$ is the natural projection. On the other hand, since V is tangent to the closed immersion $i : P/B \to G/B$, we also have a well defined vector field, say $V_e = V|_{P/B}$ on P/B. The zeros of V, V_e, and V^* are all simple isolated, and moreover

$$H^0(G/B,\mathcal{O}_Z) \cong \bigoplus_{\omega\in W} C\omega,\quad H^0(P/B,\mathcal{O}_{Z_e}) \cong \bigoplus_{\tau\in W_\Theta} C\tau,\quad H^0(G/P,\mathcal{O}_Z) \cong \bigoplus_{\hat{\omega}\in W/W_\Theta} C\hat{\omega},$$

where Z, Z_e, and Z^* are the zeros of V, V_e, and V^* respectively. We now compute the filtrations induced by these vector fields. We start first with $H^0(G/B,\mathcal{O}_Z)$.

Let α be a character of H, and let L_α be the associated homogeneous line bundle on $G/B : L_\alpha = G\mathrm{x}C/\sim$, $(g,z) \sim (g',z')$ if and only if $g' = gb$ for some $b \in B$ and $z' = \alpha(b^{-1})z$, where α is extended on B with $\alpha(u) = 1$ for u in

B_u.

Lemma: The function s_α on Z, $s_\alpha(\omega) = \omega \cdot \alpha(v)$ for $\omega \in W$, represents the first Chern class $c_1(L_\alpha^*)$ of the dual of the line bundle L_α in the isomorphism (1).

Proof: The function $\tilde{\lambda}_t : L_\alpha \to L_\alpha$, $\tilde{\lambda}_t(x,z) = (\lambda(t)x, \alpha(\lambda(t))z)$ defines a well defined C-action on L_α so that the natural map $L_\alpha \to G/B$ is C-equivariant, where $\lambda(t) = \exp(tv)$. Thus $\tilde{V} = \frac{d}{dt}(\tilde{\lambda}_t)\Big|_{t=0}$ is a V-derivation on L_α [A_2]. Since the zeros of V are all simple isolated, the function $\tilde{V}$ on Z defined by $\tilde{V}(\omega) = \frac{d}{dt}(\tilde{\lambda}_t^*(\omega x_0))\Big|_{t=0}$ represents the first Chern class $c_1(L_\alpha)$ of L_α in the isomorphism (1), where $x_0 = B \in G/B$, $\omega \in W$. Let $\lambda(t)\omega = \omega t_1$ for some $t_1 \in H$. Then we have $\tilde{\lambda}_t(\omega x_0, z) = (\omega t_1 x_0, \alpha(\lambda(t))z) = (\omega x_0, \alpha(t_1)\alpha(\lambda(t))z) = (\omega x_0, \alpha(\omega^{-1}\lambda(t)\omega)\alpha(\lambda(t))z) = (\omega x_0, \omega\alpha(\lambda(t))\alpha(\lambda(t))z)$. Thus

$$\tilde{V}(\omega) = \frac{d}{dt}(\tilde{\lambda}_t^*(\omega x_0))\Big|_{t=0} = \frac{d}{dt}(\omega\alpha(\lambda(t)^{-1})\alpha(\lambda(t)^{-1})\Big|_{t=0}$$

$$= \frac{d}{dt}(\omega\alpha(\exp(-tv))\alpha(\exp(-tv)))\Big|_{t=0} = -\omega \cdot \alpha(v) - \alpha(v).$$

Since the constant functions are in $F_0(V)$ by [A_1], [C-L_2], the function $-s_\alpha$ on Z, $-s_\alpha(\omega) = -\omega \cdot \alpha(v)$ for $\omega \in W$, represents the first Chern class $c_1(L_\alpha)$ of L_α. This proves the claim, because $c_1(L_\alpha^*) = -c_1(L_\alpha)$.

Since G is semisimple, the roots Δ of h span h^*. Thus the lemma implies that there is a well defined linear map $\psi : h^* \to F_1(V)$ determined by the condition $\psi(\alpha) = s_\alpha$ for any $\alpha \in \Delta$. Let $\psi : R \to H^0(G/B, \mathcal{O}_Z)$ ($R = \operatorname{Sym} h^*$) also denote the algebra homomorphism extending this linear map, namely $\psi(f)(\omega) = \omega \cdot f(v)$ for $f \in R$ and $\omega \in W$. Unfortunately ψ is not W-equivariant with respect to the natural action of W on $H^0(G/B, \mathcal{O}_Z)$ given by $(\sigma \cdot f)(\omega) = f(\sigma^{-1}\omega)$. To obtain equivariance, one must force W to act on $H^0(G/B, \mathcal{O}_Z)$ on the right. Thus W acts on $H^0(G/B, \mathcal{O}_Z)$ according to $(\sigma \cdot f)(\omega) = f(\omega \cdot \sigma^{-1})$. Then ψ is W-equivariant in the sense that $\sigma \cdot \psi(f) = \psi(\sigma^{-1} \cdot f)$ for $\sigma \in W$, and $f \in R$. We need only the following fact to compute the filtration $F_p(V)$ of $H^0(G/B, \mathcal{O}_Z)$.

For any $v \neq 0$ in h, let $I_v = \{f \in R^W : f(v) = 0\}$, and let $I = \{f \in R^W : f(0) = 0\}$. Then the ring R/I_vR is only graded when $I_v = I$, i.e. only when $v = 0$. However R/I_vR is filtered by degree. Namely, if $p = 0,1,\ldots$, set $(R/I_vR)_p = R_p/I_vR \cap R_p$ where $R_p = \{f \in R : \deg f \le p\}$. The natural homomorphism $\pi : R \to R/I_vR$, $\pi(f) = f(\operatorname{mod} I_vR)$ induces a surjective graded algebra homomorphism $\operatorname{gr}(\pi) : R \to \operatorname{gr}(R/I_vR) = \bigoplus_p (R/I_vR)_p/(R/I_vR)_{p-1}$. Since for $f \in I$, $f - f(v) \in I_v$ we have $\pi(I) \subset (R/I_vR)_0$, and thus $\pi(IR \cap R_p) \subset (R/I_vR)_{p-1}$. This implies $IR \subseteq \ker(\operatorname{gr}(\pi))$. But, if v is a regular vector in h, then $\dim. R/I_vR = \dim. \operatorname{gr}(R/I_vR) = |W|$. On the other hand by a theorem of Chevalley [Ch] we have $\dim R/IR = |W|$. Thus for a regular vector v in h the natural homomorphism $\pi : R \to R/I_vR$ induces a graded algebra isomorphism

$$\overline{gr(\pi)} : R/IR \xrightarrow{\sim} gr(R/I_v R) \quad) .$$

We now prove a result due to Carrell and Casselman [C].

Theorem 1. Let v be a regular vector in h. Then the algebra homomorphism $: R \to H^0(G/B,\mathcal{O}_Z)$, $\psi(f)(\omega) = \omega \cdot f(v)$, induces a W-equivariant isomorphism

$$\overline{\psi} : R/I_v R \to H^0(G/B,\mathcal{O}_Z)$$

eserving the filtrations, i.e. $\overline{\psi}((R/I_v R)_p) = F_p(V)$. Consequently for each p, $_p(V) = F_p(V)$ and the natural morphism $F_1(V)^{\otimes p} \to F_p(V)$ is surjective.

Proof: We only need to show $\ker(\psi) = I_v R$, and $\overline{\psi}((R/I_v R)_p) = F_p(V)$ for ach p. It is clear that $I_v R \subseteq \ker(\psi)$. Since ψ is W-equivariant, the variety determined by $\ker(\psi)$ is a W-invariant subvariety of $X = \{\omega \cdot v : \omega \in W\}$, the riety determined by $I_v R$. But X has no W-invariant non-trivial subset, and $\neq \phi$. Therefore $X = Y$, and thus $I_v R = \ker(\psi)$ by the Nullstellensatz, because e radical of $I_v R = I_v R$. This shows that $\overline{\psi} : R/I_v R \to H^0(G/B,\mathcal{O}_Z)$ is an isomor- ism, because $\dim R/I_v R = \dim H^0(G/B,\mathcal{O}_Z) = |W|$. On the other hand by the lemma have $\overline{\psi}((R/I_v R)_p) \subseteq F_p(v)$ for each p. Now we compare the dimensions. From e Bruhat decomposition and the isomorphism (1) one gets $\dim F_p(V) = \sum_{k=0}^{p}$ Card. $\in W : \ell(\omega) = k\}$, where $\ell(\omega)$ is the length of ω. On the other hand by the omorphism (3) we have $\dim (R/I_v R)_p = \sum_{k=0}^{p} \dim(R/IR)_k$, where $(R/IR)_k$ is the th homogeneous part of R/IR. But from the algebraic facts due to Solomon a. p. 135] and Chevalley [Ch] we get $\dim (R/IR)_k = \text{Card.}\{\omega \in W : \ell(\omega) = k\}$. Thus m $F_p(V) = \dim (R/I_v R)_p$. Since $\overline{\psi}$ is an isomorphism, we get $\overline{\psi}((R/I_v R)_p) = F_p(V)$ r each p. This finishes the proof of the theorem.

This theorem gives explicitly the filtration $F_p(V)$ of $H^0(G/B,\mathcal{O}_Z)$. By using is filtration we obtain a theorem due to Borel [B]. Let $\beta : h^* \to H^2(G/B,C)$ be e linear map determined by the condition $\beta(\alpha) = c_1(L_\alpha^*)$, the first Chern class the line bundle L_α^*, $\alpha \in X(H)$. Let $\beta : R \to H^*(G/B,C)$ be the algebra homomor- ism extending this linear map.

Corollary (Borel). The algebra homomorphism $\psi : R \to H^0(G/B,\mathcal{O}_Z)$ induces a equivariant surjective graded algebra homomorphism

$$gr(\psi) : R \to gr(H^0(G/B,\mathcal{O}_Z))$$

ch that

$$m_V \circ gr(\psi) = \beta : R \to H^*(G/B,C) .$$

reover $\ker(gr(\psi)) = IR$, and thus

$$m_V \circ \overline{gr(\psi)} = \overline{\beta} : R/IR \to H^*(G/B,C)$$

a W-equivariant graded algebra isomorphism.

Proof: It follows from the theorem 1 and the lemma because of the isomorphisms

(1) and (3).

We next compute the filtrations of $H^0(G/P,\mathcal{O}_{Z^*})$ and $H^0(P/B,\mathcal{O}_{Z_e})$. Let R^{W_Θ} be the ring of invariants of W_Θ, and let $J_v = \{f \in R^{W_\Theta} : f(v) = 0\}$, $J = \{f \in R^{W_\Theta} : f(0) = 0\}$. Since the natural map $\pi^* : H^0(G/P,\mathcal{O}_{Z^*}) \to H^0(G/B,\mathcal{O}_Z)$ is given by $\pi^*(f)(\omega) = f(\hat{\omega})$ for $\omega \in W$ and $f \in H^0(G/P,\mathcal{O}_{Z^*})$, π^* is injective, and $\pi^*(H^0(G/P,\mathcal{O}_{Z^*})) \subseteq H^0(G/B,{}_Z)^{W_\Theta}$, the ring of invariants of W_Θ. By comparing the dimensions we get $\pi^*(H^0(G/P,\mathcal{O}_{Z^*})) = H^0(G/B,\mathcal{O}_Z)^{W_\Theta}$. Consider the filtration on R^{W_Θ} given by degree. Then the algebra homomorphism $\psi_2 : R^{W_\Theta} \to H^0(G/P,\mathcal{O}_{Z^*})$, $\psi_2(f)(\hat{\omega}) = \psi(f)(\omega)$ for $f \in R^{W_\Theta}$ and $\hat{\omega} \in W/W_\Theta$, is well defined and preserves the filtrations, because π^* preserves the filtrations and the cohomology map $\pi^* : H^*(G/P,C) \to H^*(G/B,C)$ is an injection. Since $(R/I_vR)_p^{W_\Theta} = (R^{W_\Theta}/I_vR^{W_\Theta})_p$ for each p, by theorem 1 the homomorphism ψ_2 induces a filtration preserving isomorphism forming the following commutative diagram

(4)
$$\begin{array}{ccc} \bar{\psi}_2 : R^{W_\theta}/I_vR^{W_\theta} & \xrightarrow{\sim} & H^0(G/P,\mathcal{O}_{Z^*}) \\ \downarrow & & \downarrow \pi^* \\ \bar{\psi} : R/I_vR & \xrightarrow{\sim} & H^0(G/B,\mathcal{O}_Z) . \end{array}$$

Thus $F_p(V^*) = \bar{\psi}_2((R^{W_\Theta}/I_vR^{W_\Theta})_p)$ for each p.

We now compute the filtration $F_p(V_e)$ of $H^0(P/B,\mathcal{O}_{Z_e})$. Since the natural map $i^* : H^0(G/B,\mathcal{O}_Z) \to H^0(P/B,\mathcal{O}_{Z_e})$ preserves the filtrations and the cohomology map $i^* : H^*(G/B,C) \to H^*(P/B,C)$ is surjective, we have $i^*(F_p(V)) = F_p(V_e)$ for each p. This implies by theorem 1 that the algebra homomorphism $\psi_1 = i^* \circ \psi : R \to H^0(P/B,\mathcal{O}_{Z_e})$ is surjective and $\psi_1(R_p) = F_p(V_e)$ for each p. Since $J_vR \subseteq \ker(\psi_1)$ and $\dim R/J_vR = \dim H^0(P/B,\mathcal{O}_{Z_e}) = |W_\Theta|$, the order of W_Θ, the homomorphism ψ_1 induces a filtration preserving isomorphism forming the following commutative diagram

(5)
$$\begin{array}{ccc} H^0(G/B,\mathcal{O}_Z) & \xrightarrow{i^*} & H^0(P/B,\mathcal{O}_{Z_e}) \\ \bar{\psi} \uparrow \wr & & \wr \uparrow \bar{\psi}_1 \\ R/I_vR & \longrightarrow & R/J_vR \end{array}$$

Thus $F_p(V_e) = (R/J_vR)_p$ for each p.

We now summarize all of these in the following theorem. For any regular vector v in h, let $\psi : R \to H^0(G/B,\mathcal{O}_Z)$ be the algebra homomorphism given by

$(f)(\omega) = \omega \cdot f(v)$ for $f \in R$ and $\omega \in W$. If $\psi_1 : R \to H^0(P/B, \mathcal{O}_{Z_e})$, $\psi_2 : R^{W_\Theta} \to$ $^0(G/P, \mathcal{O}_{Z^*})$ are the algebra homomorphisms given by $\psi_1(f)(\tau) = \psi(f)(\tau)$, $\psi_2(g)(\hat{\omega}) =$ $(g)(\omega)$, then we have:

Theorem 2. The following diagram of graded algebras

$$\begin{array}{ccccc}
H^*(G/P,C) & \xrightarrow{\pi^*} & H^*(G/B,C) & \xrightarrow{i^*} & H^*(P/B,C) \\
\uparrow \wr \; m_{V^*} & & \uparrow \wr \; m_V & & \uparrow \wr \; m_{V_e} \\
gr(H^0(G/P,\mathcal{O}_{Z^*})) & \longrightarrow & gr(H^0(G/B,\mathcal{O}_Z)) & \longrightarrow & gr(H^0(P/B,\mathcal{O}_{Z_e})) \\
\uparrow \wr \; \overline{gr(\psi_2)} & & \uparrow \wr \; \overline{gr(\psi)} & & \uparrow \wr \; \overline{gr(\psi_1)} \\
R^{W_\Theta}/IR^{W_\Theta} & \longrightarrow & R/IR & \longrightarrow & R/JR
\end{array}$$

s commutative.

Proof. The isomorphism $\overline{gr(\psi_2)}$ follows diagram (4) and theorem 1, because $R/I_vR)^{W_\Theta}_p = (R^{W_\Theta}/I_vR^{W_\Theta})_p$ for each p. On the other hand, since W_Θ is a reflec-ion group, by a similar argument given in the proof of isomorphism (3) we get $r(\psi_1) : R/JR \cong gr(H^0(P/B,\mathcal{O}_{Z_e}))$, because $\overline{\psi}_1 : R/J_vR \cong H^0(P/B,\mathcal{O}_{Z_e})$ is a filtration reserving isomorphism. Thus the rest follows from theorem 1, and diagrams (2),), (5).

We now prove a result similar to corollary of theorem 1. Let $S = C[\Theta] \subset R$ e the polynomial algebra in the variables $\alpha \in \Theta$, and let $\beta_\Theta : S \to H^*(P/B,C)$ be ie algebra homomorphism determined by $\beta_\Theta(\alpha) = c_1(L^*_\alpha)$, the first Chern class of ie dual of the line bundle L_α on P/B, for any $\alpha \in \Theta$. S is invariant under). If we set $I^\Theta = \{f \in S^{W_\Theta} : f(0) = 0\}$, where S^{W_Θ} is the ring of invariants W_Θ, then we have:

Corollary. The inclusion $i : S \to R$ induces an isomorphism

$$\overline{i} : S/I^\Theta S \to R/JR$$

) that

$$m_{V_e} \circ \overline{gr(\psi_1)} \circ \overline{i} = \overline{\beta}_\Theta : S/I^\Theta S \cong H^*(P/B,C) .$$

Proof. Everything follows from the theorem above and corollary of theorem 1 xcept the isomorphism $\overline{i} : S/I^\Theta S \to R/JR$. We now show that $\overline{i}$ is an isomorphism.) see this, let $h_1 \subset h$ be the dual space of the space spanned by Θ in h^*, id v_1 be a regular vector in h_1 for W_Θ. Consider the vector field V_1 on $/\widetilde{B}$ induced from the one parameter family $\exp(tv_1)$, where $\widetilde{P} = P/R(P)$,

$\tilde{B} = B/R(P)$, $R(P)$ is the radical of P. Then by theorem 1 ($\tilde{P}$ is semisimple) and diagram (5) the natural isomorphism $p : P/B \overset{\sim}{\to} \tilde{P}/\tilde{B}$ induces a filtration preserving isomorphism forming the following commutative diagram

$$\begin{array}{ccc} H^0(\tilde{P}/\tilde{B}, \mathcal{O}_{Z_1}) & \xrightarrow{\ \sim\ } & H^0(P/B, \mathcal{O}_{Z_e}) \\ \uparrow \wr & & \wr \uparrow \bar{\psi}_1 \\ S/I^{\Theta}_{v_1} S & \xrightarrow{\ \bar{i}\ } & R/J_v R \end{array},$$

where $Z_1 = \mathrm{zero}(V_1)$, $I^{\Theta}_{v_1} = \{f \in S^{W_\Theta} : f(v_1) = 0\}$, and $\bar{i} : S/I^{\Theta}_{v_1} S \to R/J_v R$ is the homomorphism induced from the inclusion $i : S \to R$ (here v is considered as $v = v_1 + v_2$ for some $v_2 \in h$). But this implies that $\bar{i} : S/I^{\Theta}_{v_1} S \to R/J_v R$ is a filtration preserving isomorphism. Thus the associated graded algebra homomorphism $gr(\bar{i}) : gr(S/I^{\Theta}_{v_1} S) \to gr(R/J_v R)$ gives, by theorem 2, diagram (5), and the isomorphism (3), the isomorphism $\bar{i} : S/I^{\Theta} S \to R/JR$.

Remark:

(1) Here we have computed the filtration of $H^0(G/P, \mathcal{O}_Z)$ for the regular vector fields. At the other extreme G/P always admits a vector field with exactly one zero by [A_3]. Recently, it is shown by [A-C-L-S] that there exists a unique vector field V with one zero on G/P admitting a C^*-action, and the filtration of $H^0(G/P, \mathcal{O}_Z)$ for this vector field is given by the height of the roots. This description of $H^0(G/P, \mathcal{O}_Z)$ gives information about the Schubert calculus like the one given in [B-G-G] by means of [A_4] and [A_5].

(2) Theorem 1 and its corollary have also been obtained by J.B. Carrell and W. Casselman. But their proof was only an outline.

REFERENCES

[A_1] Akyildiz, E., Ph.D. Thesis, Univ. of British Columbia, Vancouver, B.C., 1977.

[A_2] __________, Vector fields and Equivariant bundles, Pac. Jour. of Math. 81 (1979), 283-289.

[A_3] __________, A vector field with one zero on G/P, Proc. Amer. Math. Soc. 67 (1977), 32-34.

[A_4] __________, Bruhat decomposition via G_m-action, Bull. Acad. Pol. Sci., Sér. Sci., Math,. 28 (1980), 541-547.

[A_5] __________, On the G_m-decomposition of G/P, METU Jour. Pure and Applied Sci., to appear.

[$A-C_1$] Akyildiz, E. and Carrell, J.B., Zeros of holomorphic vector fields and the Gysin homomorphism, Proc. Symp. Pure Math. Summer Inst. on Singularities (Arcata), to appear.

[$A-C_2$] Akyildiz, E. and Carrell, J.B., Gysin homomorphism of homogeneous spaces, submitted.

[A-C-L-S] Akyildiz, E. and Carrell, J.B. and Lieberman, D.I. and Sommese, A.J., On the graded rings associated to holomorphic vector fields, to appear.

[B-G-G] Bernstein, I.N. and Gel'fand, I.M. and Gel'fand, S.I., Schubert cells and cohomology of the space G/P, Russian Math. Surveys 28 (1973), 1-26.

[B] Borel, A., Sur la cohomologie des espaces fibrés principaux et des espaces homogenes des groupes de Lie compacts, Ann. of Math (2) 57 (1953), 115-207.

[C] Carrell, J.B., Vector fields and the cohomology of G/B, to appear.

[$C-L_1$] Carrell, J.B. and Lieberman, D.I., Holomorphic vector fields and Kaehler manifolds, Invent. Math 21 (1974), 303-309.

[$C-L_2$] Carrell, J.B. and Lieberman, D.I., Vector fields and Chern numbers, Math. Ann. 225 (1977), 263-273.

[Ca] Carter, R.W., Simple groups of Lie type, John Wiley and Sons, New York, 1972.

[Ch] Chevalley, C., Invariants of finite groups generated by reflections, Amer. J. Math. 67 (1955), 778-782.

[G-H] Griffiths, P. and Harris, H., Principles of Algebraic geometry, John Wiley and Sons, New York, 1978.

[H] Humphreys, J.E., Linear algebraic groups, Springer-Verlag, Berlin and New York, 1975.

Middle East Technical University
Department of Mathematics
Ankara, Turkey.

COMPLETE QUOTIENTS BY ALGEBRAIC TORUS ACTIONS

by

Andrzej Bialynicki-Birula and Joanna Swiecicka

The aim of the paper is to provide an answer to the following problem:

Let an algebraic torus T act on a normal complete variety X. Describe all open T-invariant subsets U of X for which the geometric quotient space U/T exists and is complete.

The problem has been studied in [B-B,S] for algebraic actions defined over the complex number field $\mathbb{C}$ and for complex analytic actions. Here we consider the algebraic case where the ground field k is algebraically closed of any characteristic. The main result is in the spirit of [B-B,S], however our proof is based on completely different ideas. The proof gives in fact a result concerning more general quotients. The answer to the problem starting at the beginning follows directly from this result.

§1. Notations and Terminology. The main result.

The ground field k is assumed to be algebraically closed; all algebraic varieties and their morphisms are supposed to be defined over k. Let T denote a one-dimensional torus and let X be a normal complete algebraic variety. Assume that we have an action of T on X. For any $t \in T$ and $x \in X$, tx denotes the value at x of the automorphism of X assigned to t.

Let $X^T = X_1 \cup \cdots \cup X_r$ be the decomposition of the fixed point set of the action given on X into connected components.

Let us fix an isomorphism $T \simeq k^*$. Then for any $x \in X$, the morphism $\phi_x : T \to X$, defined by $\phi_x(t) = tx$, can be extended to the projective line $P^1(k) \supset k^*$. The extended morphism will be also denoted by ϕ_x. Define

$$\Phi^+(x) = \phi_x(0), \quad \Phi^-(x) = \phi_x(\infty)$$

$$X_i^+ = \{x \in X;\ \Phi^+(x) \in X_i\}, \quad X_i^- = \{x \in X,\ \Phi^-(x) \in X_i\} .$$

Definition 1.1. Let $i, j \in \{1,2,\ldots,r\}$. We say that X_i is directly less than X_j if there exists $x \in X - X^T$ such that $\Phi^+(x) \in X_i$, $\Phi^-(x) \in X_j$. We say that X_i is less than X_j and we write $X_i < X_j$ if there exists a sequence $i = i_0, i_1, \ldots, i_\ell = j$ such that $X_{i_{s-1}}$ is directly less than X_{i_s}, for $s = 1,\ldots,\ell$. We shall write $X_i \le X_j$ if $X_i < X_j$ or $X_i = X_j$.

Definition 1.2. A semi-section of $\{1,2,\ldots,r\}$ is a division of $\{1,\ldots,r\}$ into three disjoint subsets A^+, A^0, A^- satisfying the following condition:

$$\text{if } i \in A^+ \cup A^0 \text{ and } X_j < X_i \text{ then } j \in A^+ .$$

A section of $\{1,2,\ldots,r\}$ is a semi-section (A^+,A^0,A^-) where $A^0 = \emptyset$.

Definition 1.3. Let (A^+,A^0,A^-) be a semi-section of $\{1,2,\ldots,r\}$ and let $U = \bigcup_{\substack{i\in A^+\cup A^0 \\ j\in A^-\cup A^0}} (X_i^+ \cap X_j^-)$. Then U is called a semi-sectional set corresponding to the semi-section (A^+,A^0,A^-) . A semi-sectional set corresponding to a section is called a sectional set.

Notice, that if U is a semi-sectional set corresponding to a semi-section (A^+,A^0,A^-) then

$$U = X - \Big(\bigcup_{i\in A^+} X_i^- \cup \bigcup_{i\in A^-} X_i^+\Big)$$

In this paper we are going to consider two concepts of quotient maps: a geometric quotient and a semi-geometric quotient. The notion of a geometric quotient was introduced by Mumford in [G.I.T.]. In the special case, we are considering in this paper, his definition is equivalent to the following:

Definition 1.4. Let an algebraic torus T act on an algebraic variety X . A morphism $\pi: X \to Y$, where Y is an algebraic variety, is said to be a geometric quotient of X (with respect to the given action of T) if the following conditions are satisfied:

(a) for any $y \in Y$, $\pi^{-1}(y)$ is an orbit in X ,

(b) π is an affine morphism,

(c) for any open affine $U \subset Y$, the ring $k[U]$ of regular functions on U is identified by π^* with the ring $k[\pi^{-1}(U)]^T$ of regular T-invariant functions on $\pi^{-1}(U)$.

Definition 1.5. Let X, Y, π be as in Definition 1.4. The morphism $\pi: X \to Y$ is said to be a semi-geometric quotient of X (with respect to the given action of T) if conditions (b) and (c) of Definition 1.4 are satisfied. (The notion of a semi-geometric quotient is equivalent to the notion of a good quotient of Seshadri, see [Se])

We are going to denote a semi-geometric quotient of X by $X \to X/T$.

It is easy to see that if a semi-geometric quotient exists, then it is a categorical quotient. On the other hand if $\pi: X \to Y$ is a categorical quotient and π is an affine morphism, then $\pi: X \to Y$ is a semi-geometric quotient.

Now, we are ready to state the main result of the paper.

Theorem. Let X be a normal complete algebraic variety with an action of T . If U is a semi-sectional subset of X , then U is open, T-invariant, a semi-geometric quotient $U \to U/T$ exists, and U/T is complete. Conversely, if U is an open, T-invariant subset of X such that a semi-geometric quotient $U \to U/T$ exists and U/T is complete, then U is a semi-sectional subset of X.

Corollary. Let X be as in the theorem. If U is sectional, then a geometric quotient $U \to U/T$ exists and U/T is complete. If for an open, T-invariant subset $U \subset X$, a geometric quotient exists with U/T complete, then U is sectional.

In fact the above theorem summarizes the contents of Theorem 3.1 and Theorem 3.3 and the corollary follows directly from Remark 3.9 proved in §3.

Remark 1.6. It should be noted that for actions of T on smooth projective varieties one can find such semi-sectional sets U which have complete non-projective quotient spaces U/T. For example if X = Grassmannian (2,4) with the action of T induced by the action on k^4 given by matrices

$$\begin{pmatrix} t^{n_1} & & & 0 \\ & t^{n_2} & & \\ 0 & & t^{n_3} & \\ & & & t^{n_4} \end{pmatrix}, \text{ where } n_1, n_2, n_3, n_4 \text{ are pairwise different integers,}$$

then for at least one (and at most two depending on the choice of the integers n_1, n_2, n_3, n_4) sectional sets, the quotient space is not projective.

Remark 1.7. If U is a semi-sectional subset of U, then U is a maximal subset of X which is open, T-invariant and which admits a semi-geometric quotient. However, in general, for a normal complete algebraic variety X the semi-sectional subsets are not the only maximal subsets with the above properties (see [B-B,S]).

§2. Auxiliary results.

If L is a T-linearized invertible sheaf on X, then X^{ss}_L or $X^{ss}(L)$ denotes the set of semi-stable points of X with respect to L (see [G.I.T.] for definitions). It has been proved by Mumford [G.I.T], that X^{ss}_L is open, T-invariant and that the semi-geometric quotient $X^{ss}_L \to X^{ss}_L/T$ exists and X^{ss}_L/T is quasi-projective. On the other hand, if for some algebraic variety U the semi-geometric quotient $U \to U/T$ exists and U/T is quasi-projective, then $U = U^{ss}_L$ for some T-linearized invertible ample sheaf L on U.

Example 2.1. Suppose $T = k^*$ acts on an n-dimensional projective space $\mathbb{P}$. Then we can find a coordinate system on $\mathbb{P}$ such that the action of k^* is given by

$$t[x_0,\dots,x_n] = [x_0,\dots,x_n]\begin{pmatrix} t^{n_1} & & 0 \\ & \ddots & \\ 0 & & t^{n_n}\end{pmatrix}, \text{ for any } t \in k^* \text{ and } [x_0,\dots,x_n] \in \mathbb{P},$$

where $0 = n_1 = \dots = n_{i_1} < n_{i_1+1} = \dots = n_{i_2} < \dots\dots < n_{i_{r+1}} = \dots = n_n$ is a sequence of integers.

Then $\mathbb{P}^T = P_1 \cup\dots\cup P_r$, where $P_i = \{[x_0,\dots,x_n] \in \mathbb{P};\ x_0 = \dots = x_{i_{r-1}} = 0 = x_{i_{r+1}} = \dots = x_n\}$.

For any $a = [a_0,\dots,a_n] \in \mathbb{P}$, let a_ℓ be the non-zero coordinate of a with the smallest index and let a_k be the non-zero coordinate of a with the greatest index. Let $i_{i-1} < \ell \le i_i$ and $i_{j-1} < k < i_j$. Then $\Phi^+(x) \in \mathbb{P}_i$, $\Phi^-(x) \in P_j$. Hence for any i,j, $i < j$, P_i is directly less than P_j, in particular $P_i < P_j$. Moreover if $P_i \le P_j$ and $P_j \le P_i$, then $P_i = P_j$.

If X is a projective variety with an action of T embedded in an equivariant way into $\mathbb{P}$ (any normal quasi-projective variety with an action of T has such an embedding, see [S]), then it follows from the above that $X_i \le X_j$ and $X_j \le X_i$ implies $X_i = X_j$.

All sections (A^+,A^0,A^-) of $\{1,\dots,r\}$ (for the given action of T on $\mathbb{P}$) are of the form $A^+ = \{1,\dots,i\}$, $A^0 = \emptyset$, $A^- = \{i+1,\dots,r\}$, all semi-sections which are not sections are of the form $A^+ = \{1,\dots i\}$, $A^0 = \{i+1\}$, $A^- = \{i+2,\dots,r\}$.

All semi-sectional subsets of $\mathbb{P}$ are open. Moreover it follows directly from [GIT] that for any semi-sectional set $U \subset \mathbb{P}$ there exists a T-linearization of $\theta(1)$ such that $U = P^{ss}(\theta(1))$, hence a semi-geometric quotient $U \to U/T$ exists with U/T projective. The quotient is geometric, if U is sectional.

<u>Lemma 2.2</u>. Let L be a T-linearized ample invertible sheaf on a projective variety X. Then the set X_L^{ss} is semi-sectional.

<u>Proof</u>. Replacing L by $L^{\otimes n}$, for some natural number n, if necessary, we may assume that L is very ample. Let $\phi_L\colon X \to \mathbb{P}^m$ be an embedding determined by L. The given T-linearization of L gives a representation of T in the space of global sections $\Gamma(X,L)$ and hence it gives a T-linearization of $\theta(1)$ on $\mathbb{P}^m$ (we shall call it a T-linearization of $\theta(1)$ corresponding to the given linearization of L). The set $(P^m)^{ss}(\theta(1))$ is semisectional (Example 2.1). Moreover, for any global section s of L the set $\{x \in X;\ s(x) \neq 0\}$ is affine (since L is very ample) and any T-invariant global section of L can be extended to a T-invariant section of $\mathcal{O}(1)$ (we identify X and $\phi_L(x) \subset \mathbb{P}^m$). The set where the extended section is different from zero is again affine. This shows that $X^{ss}(L) \subset (\mathbb{P}^m)^{ss}(\mathcal{O}(1))$. Similarly $(\mathbb{P}^m)^{ss}(\mathcal{O}(1)) \cap X \supset X^{ss}(L)$. Hence

$(\mathbb{P}^m)^{ss}(\mathcal{O}(1)) \cap X = X^{ss}(\mathcal{L})$. Since $(\mathbb{P}^m)^{ss}(\mathcal{O}(1))$ is semi-sectional (Example 2.1), $X_{\mathcal{L}}^{ss}$ is also semi-sectional.

Proposition 2.3. Let X be a (not necessarily normal) algebraic complete variety with an action of an algebraic torus T. Let $X^T = X_1 \cup \ldots \cup X_r$ be the decomposition into connected compoments, where X_1 is the source and X_r is the sink. Then $X_1 \leq X_j \leq X_r$, for any $j = 1,\ldots,r$.

Proof. Assume first that X is normal and projective. Suppose that X_1 is not $\leq X_i$, for some i, and let X_j be a minimal (with respect to $<$) component with this property. Then $X_j^- = \emptyset$ or there exists $X_k \neq X_j$ such that $X_k < X_j$ and (by the minimality assumption) $X_j < X_k$. If $X_j^- = \emptyset$ then X_j is the source X_1, contradicting with the assumption that X_1 is not $\leq X_j$. The second possibility can not occur either because X can be T-equivariantly embedded in $\mathbb{P}^m$ and we may use the result stated in Example 2.1.

Now, if X is any complete algebraic variety then there exists a normalization morphism $\eta: \tilde{X} \to X$ and by an Equivariant Chow Lemma (Theorem 2 [Su]) a birational T-equivariant morphism $\theta: \tilde{\tilde{X}} \to \tilde{X}$, where $\tilde{\tilde{X}}$ is projective and normal. For a connected component $X_i \subset X$, let Y be any connected component of $\tilde{\tilde{X}}^T$ contained in $(\eta\theta)^{-1}(X_i)$. Then by the first part of the proof the source of $\tilde{\tilde{X}}$ is less than Y and hence there exists a sequence of points $x_1,\ldots,x_s \in \tilde{\tilde{X}}$ such that $\Phi^+(x_1)$ belongs to the source of $\tilde{\tilde{X}}$, $\Phi^-(x_i)$ and $\Phi^+(x_{i+1})$ belong to the same connected component of $\tilde{\tilde{X}}^T$ for $i = 1,\ldots,s-1$, and $\Phi^-(x_s) \in Y$. Then the sequence $\eta\theta(x_1),\ldots,\eta\theta(x_s)$ has the following properties: $\Phi^+(\eta\theta(x_1)) \in X_1$, $\Phi^-(\eta\theta(x_i))$ and $\Phi^+(\eta\theta(x_{i+1}))$ belong to the same connected component of X^T and $\Phi^-(\eta\theta(x_s)) \in X_j$. Thus $X_1 \leq X_i$. By symmetry $X_i \leq X_r$.

Proposition 2.4. Any semi-sectional subset $U \subset X$ is open in X.

Proof. Let U correspond to the semi-section (A^+,A^0,A^-). Then $U = X - (\bigcup_{i,j\in A^+} (X_i^+ \cap X_j^-) \cup \bigcup_{i,j\in A^-} (X_i^+ \cap X_j^-))$. We shall prove that $\bigcup_{i,j\in A^+}(X_i^+ \cap X_j^-)$, $\bigcup_{i,j\in A^-}(X_i^+ \cap X_j^-)$ are closed. Let $i,j \in A^+$, let B be an irreducible component of $\overline{X_i^+ \cap X_j^-}$, let $B_1,\ldots,B_s$ be the connected components of B^T, with B_1-the source and B_s-the sink. Let $x \in B$ and let $\Phi^+(x) \in B_k$, $\Phi^-(x) \in B_\ell$. Then by Proposition 2.3. $B_1 \leq B_k \leq B_\ell \leq B_s$. Let $B_k \subset X_r$, $B_\ell \subset X_t$. Then $X_i \leq X_r \leq X_t \leq X_j$ and $x \in X_r^+ \cap X_t^-$. Thus $x \in \bigcup_{i,j\in A^+}(X_i^+ \cap X_j^-)$ and thus $\bigcup_{i,j\in A^+}(X_i^+ \cap X_j^-)$ is closed. By symmetry $\bigcup_{i,j\in A^-}(X_i^+ \cap X_j^-)$ is closed. Hence U is open.

Proposition 2.5. Let U be a semi-sectional set corresponding to the semi-section

(A^+, A^0, A^{-1}) and let $\phi: U \to U/T$ be a semi-geometric quotient. Then $V = U - \bigcup_{i\in A^0}(X_i^+ \cup X_i^-)$ is open and $\phi|V: V \to \phi(V)$ is a geometric quotient.

Proof follows from Proposition 2.3, Definitions 1.4 and 1.5, Amplification 1.3 p. 30 [GIT] and the remark that all orbits contained in V are closed in U.

In the sequel we shall use the following well known

Lemma 2.6. Let $U \to U/T$ be a semi-geometric quotient. If U is normal then U/T is also normal.

Finally, we shall need the following two lemmas on gluing of algebraic varieties.

Lemma 2.7. Suppose that we have the following commutative diagram of morphisms of algebraic varieties (where $\hookrightarrow$ denotes open immersion)

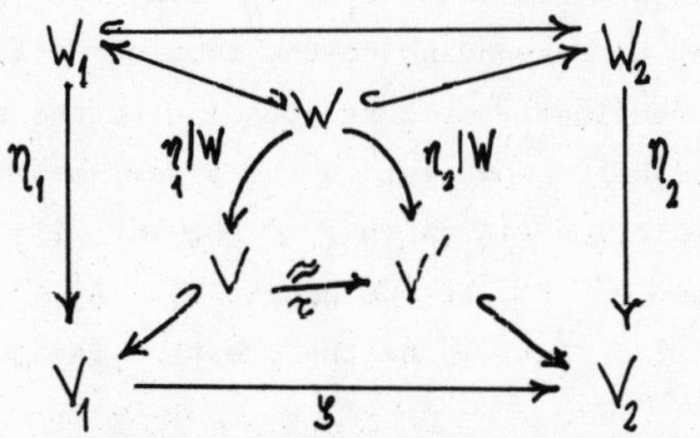

Assume that τ is an isomorphism and that V_2 and $\eta_1(W_1-W)$ are complete and $\zeta\eta_1(W_1-W) = \eta_2(W_2-W)$. Then $V = (V_2 - \eta_2(W_2-W)) \cup_\tau V_1$ is a complete algebraic variety and the induced map

$$\eta: (W_2 - \eta_2^{-1}(\eta_2(W_2-W_1))) \cup W_1 \to V \text{ is a morphism.}$$

The proof follows from the valuative criteria of properness and separatedness.

Similarly one proves the following

Lemma 2.8. Suppose that we have the following commutative diagram of morphisms of algebraic varieties

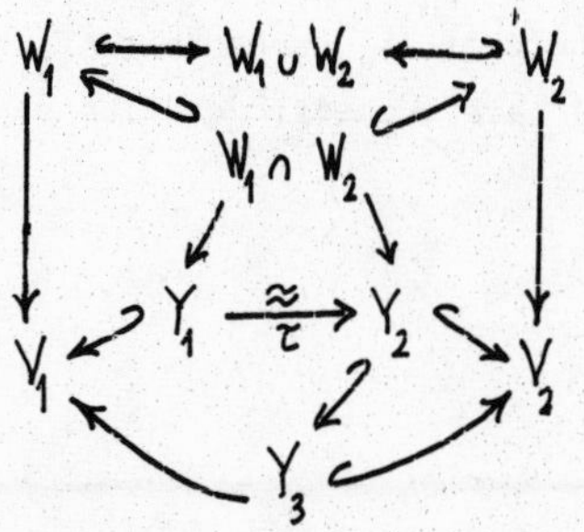

Assume that ζ is surjective, V_2 and $Y_3 - Y_2$ are complete.

Then $V = (V_2 - (Y_3-Y_2)) \cup_\tau V_1$ is a complete algebraic variety and the induced map $\eta: W_1 \cup W_2 \twoheadrightarrow V$ is a morphism.

§3. Main results.

Theorem 3.1. Let X be normal complete variety and let T act on X. For any semi-sectional $U \subset X$, U is open and T-invariant, the semi-geometric quotient $U \twoheadrightarrow U/T$ exists and U/T is complete.

Proof. It follows from Proposition 2.4 that U is open, and from Definition 1.3 that U is T-invariant.

Now, assume that X is projective. Let $\psi: X \to \mathbb{P}^n$ be a T-equivariant embedding of X into a projective space $\mathbb{P}^n$. Let $\mathbb{P}_1,\ldots,\mathbb{P}_k$ be the irreducible components of $(\mathbb{P}^n)^T$ and suppose that $P_i < P_j$ for $i < j$; $i,j = 1,\ldots,k$. Let U_{2i} be the sectional set corresponding to the section $(\{1,\ldots,i\},\emptyset,\{i+1,\ldots,k\})$ and U_{2i-1} be the semi-sectional set corresponding to the semi-section $(\{1,\ldots,i-1\},\{i\},\{i+1,\ldots,k\})$. Now, let U be a semi-sectional subset of X corresponding to a semi-section (A^+,A^0,A^-). For any $i = 1,\ldots,2k-1$, $U_i \cap \psi(X)$ is a semi-sectional subset of X, let us denote by (A_i^+,A_i^0,A_i^-) the corresponding semi-section of $\{1,\ldots,r\}$. Let m be the greatest integer such that $A_m^+ \subset A^+$, $A_m^0 \subset A^0$.

We shall prove existence of the semi-geometric quotient $\phi: U \to U/T$ and completeness of U/T by induction on $\ell(U) = 2\#(A^+-A_m^+) + \#(A^0-A_m^0)$. If $\ell(U) = 0$ then $U = \psi^{-1}(U_m)$ and the semi-geometric quotient exists and U/T is complete since it exists for U_m and U_m/T is projective (see Example 2.1). Assume that the theorem is proved for all semi-sectional subsets V with $\ell(V) < \ell$ and let, for U, $\ell(U) = \ell > 0$. Let n be the smallest integer such that $A^+ \subset A_n^+$, $A^0 \subset A_n^0$. Then $n > m$.

Case (a). $A_n^0 \neq \emptyset$. Then $A^0 \neq \emptyset$. Let $j \in A^0$. Then $U' = U - X_j^-$ corresponds to the semi-section $(A^+,A^0-\{j\}, A^- \cup \{j\})$ and $\ell(U') = \ell(U) - 1 < \ell$. Thus a semi-geometric quotient $\phi': U' \to U'/T$ exists and U'/T is complete. On the other hand a semi-geometric quotient $\phi_n: U_n \to U_n/T$ exists with U_n/T projective. Let V_j be a T-invariant neighbourhood of X_j such that $V_j \subset U_n \cap U$ and $\phi_n|(V_j - (X_j^+ \cup X_j^-))$, $\phi'|(V_j \cap U')$ are geometric quotients. Then we have the following commutative diagram

$$U = V_j \cup (U' - X_j^-) = U \cup U'$$

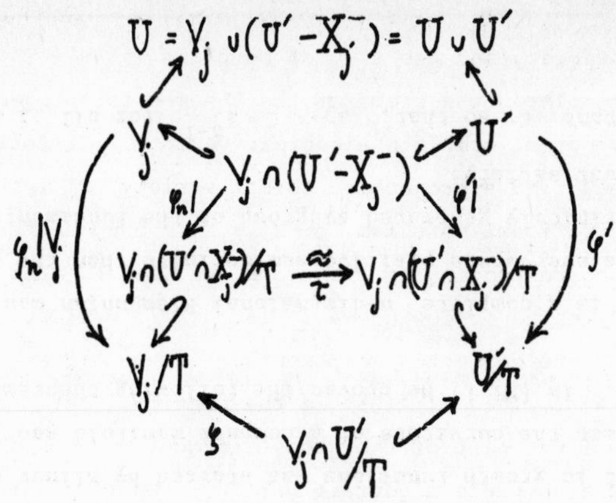

where ζ is the morphism of quotients induced by open embedding $V_j \cap U' \subset V_j$, and τ is the canonical isomorphism of geometric quotients. Let $V = V_j/T \cup_\tau ((U'-X_j^-)/T)$. By Lemma 2.8 V is a complete algebraic variety and the induced morphism $U \to V$ is a semi-geometric quotient.

Case (b). $A_n^0 = \emptyset$. Let $j \in (A_n^+ - A_{n-1}^+) \cap A^+$ (such j exists since n is the smallest integer for which $A_n^+ \supset A^+$). Let U' correspond to the semi-section $(A^+ - \{j\}, \{j\}, A^-)$. Then $U' = U \cup X_j^+$. Moreover $\ell(U') = \ell(U) - 1 < \ell$ and hence a semi-geometric quotient $\phi': U' \to U'/T$ exists and U'/T is complete. Let V_j be a T-invariant neighbourhood of X_j such that $V_j \subset U' \cap U_{n+1}$ and $\phi'|V_j-X_j^+$ is a geometric quotient.

Then we have the following commutative diagram:

$$
\begin{array}{ccc}
V_j - X_j^+ & \longhookrightarrow & U' = U \cup X_j^+ \\
& V_j - (X_j^+ \cup X_j^-) & \\
\zeta \downarrow & \phi_n \downarrow & \downarrow \phi' \\
V_j - (X_j^+ \cup X_j^-)/T & \xrightarrow{\approx}_\tau & V_j - (X_j^+ \cup X_j^-)/T \\
V_j - X_j^+/T & \xrightarrow{\zeta} & U'/T
\end{array}
$$

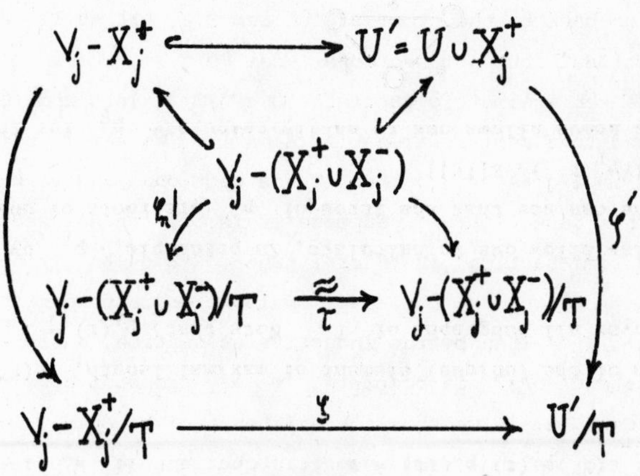

where τ is the canonical isomorphism of geometric quotients and ζ is induced by $V_j - X_j^+ \hookrightarrow U'$. Let $V = (U'/T - \phi'(X_j^+ \cup X_j^-)) \cup_\tau V_j - X_j^+/T$. Then by Lemma 2.7 V is a complete algebraic variety and the induced morphism $\phi: U \to V$ is a semi-geometric quotient. This completes the proof of the theorem for projective X.

Now, let X be any normal complete algebraic variety. It follows from Theorem 2 [S], that we may find a projective normal variety X' with an action of T and a T-equivariant birational morphism $\xi: X' \to X$. Suppose that U is a sectional subset of X corresponding to a section $(A^+, \emptyset, A^-)$. Let $(B^+, \emptyset, B^-)$ be a section for X' defined in the following way: let $(X')^T = X'_1 \cup \ldots \cup X'_s$ be the decomposition of $(X')^T$ into connected components; $i \in B^+$ if and only if $\xi(X'_i) \subset X_j$, where $j \in A^+$. Then $\xi^{-1}(U)$ is the sectional set corresponding to $(B^+, \emptyset, B^-)$. Thus by the first part of the geometric quotient $\xi^{-1}(U) \to \xi^{-1}(U)/T$ exists and $\xi^{-1}(U)/T$ is a complete algebraic variety. On the other hand we have $U \to U/T$ where U is an algebraic prevariety. ξ induces a surjective birational morphism $\xi^*: \xi^{-1}(U)/T \to U/T$. In order to prove that U/T is separated it suffices to show the following:

Lemma 3.2. Let $\xi^*: W_1 \to W_2$ be a surjective morphism of a complete algebraic variety W_1 onto an algebraic prevariety W_2. Assume that $\xi^*|(\xi^*)^{-1}(V): (\xi^*)^{-1}(V) \to V$ is an isomorphism for some dense and open subset $V \subset W_2$. Then W_2 is a complete variety.

Proof. Since W_1 is complete, if W_2 is separated, then W_2 is a complete algebraic variety. Let $\mathcal{O}$ be a valuation ring in the field $k(W_2) = k(V) = k(W_1)$. Suppose that $\mathcal{O}$ dominates $\mathcal{O}_{x_1}, \mathcal{O}_{x_2}$ where $x_1, x_2 \in W_2$. Then $\mathcal{O}$ dominates $\mathcal{O}_{y_1}, \mathcal{O}_{y_2}$, for some points $y_1, y_2 \in W_1$ such that $\xi^*(y_1) = x_1$, $\xi^*(y_2) = x_2$ (because ξ^* is proper). Since W_1 is separated, $y_1 = y_2$. Thus $x_1 = x_2$. The proof of lemma is finished.

Let us now go back to the proof of Theorem 3.1. It still remains to show that for any semi-sectional (but not sectional) set $U \subset X$, where X is a complete normal variety, a semi-geometric quotient $U \to U/T$ exists and U/T is complete. Let U correspond to a semi-section (A^+, A^0, A^-). We shall prove this result by induction on $\#A^0$. If $\#A^0 = 0$, then U is sectional and we know already that the theorem in this case is true. Suppose it is true when a semi-sectional set corresponds to a semi-section (B^+, B^0, B^-) where $\#B^0 < \#A^0$. Suppose $\#A^0 > 0$ and let $j \in A^0$. Let U' be the semi-sectional set corresponding to the semi-section $(A^+, A^0 - \{j\}, A^- \cup \{j\})$. Then by the inductive assumption, a semi-geometric quotient $U' \to U'/T$ exists and U'/T is complete. We may proceed, now, in the same way as in the proof of Case (a) if we show the existence of an open T-invariant neighbourhood V_j of X_j for which a semi-geometric quotient $\phi_j: V_j \to V_j/T$ exists (and

$\phi_j|V_j - (X_j^+ \cup X_j^-)$ is a geometric quotient). This follows from the existence of a quasi-projective, T-invariant neighbourhood of any point $x \in X_j$, the embedding theorem of Sumihiro [S] and the existence of semi-geometric quotients of semi-sectional subsets of $\mathbb{P}^n$ (Example 2.1). The proof of Theorem 3.1 is complete.

Theorem 3.3. Let X be a normal complete algebraic variety with an action of T. Let $U \subset X$ be an open T-invariant subset of X such that the semi-geometric quotient $\phi: U \to U/T$ exists and U/T is complete. Then U is a semi-sectional set in X.

The plan of the proof is the following. First we prove.

Proposition 3.4. Let U be an algebraic normal variety with an action of T for which a semi-geometric quotient $U \to U/T$ exists and U/T is complete. Then there exists a normal variety U' with an action of T and a birational proper T-equivariant morphism $\Phi: U' \to U$ such that the semi-geometric quotient $U' \to U'/T$ exists and U'/T is projective.

In the next step we shall find a good equivariant embedding of U' in a normal projective variety; more exactly we shall prove the following

Proposition 3.5. Let U' be an algebraic normal variety with an action of T for which the semi-geometric quotient $U' \to U'/T$ exists and U'/T is projective. Then there exist a noraml projective variety X' with an action of T and an open T-equivariant embedding $i: U' \hookrightarrow X'$ such that $i(U')$ is contained in a semi-sectional set W for which the semi-geometric quotient W/T is projective.

Then we shall prove that $i(U')$ is in fact a semi-sectional set. We shall identify, in the sequel, U' and $i(U')$.

Proposition 3.6. Let U' be an open T-invariant subset of a normal projective variety X' such that the semi-geometric quotient $U' \to U'/T$ exists with U'/T complete. If U' is contained in a semi-sectional subset $W \subset X'$, then U' is equal to some semi-sectional set in X'.

Let ϕ' be a birational map $X' \dashrightarrow X$ defined by a birational morphism of open subsets $\Phi: U' \to U$. Taking the closure of the graph of ϕ' in $X \times X'$ we find an algebraic variety $\tilde{X}$ with an action of T and T-equivariant morphisms $\pi: \tilde{X} \to X'$ and $\tau: \tilde{X} \to X$ such that $\phi'\pi = \tau$.

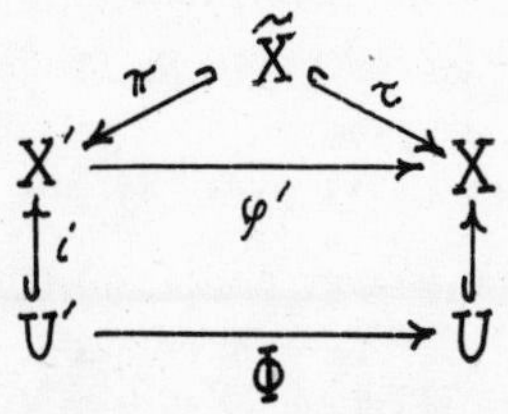

Note that $\pi^{-1}(i(U')) \simeq U'$ and $\tau^{-1}(U) = \pi^{-1}(i(U'))$. The latter follows from the fact that $\tau(\pi^{-1}(i(U'))) = U$ and $\tau|\pi^{-1}(i(U')) = \phi'\pi|\pi^{-1}(i(U'))$ is proper. To complete the proof of the theorem it suffices to prove the following two propositions:

<u>Proposition 3.7</u>. Let $\pi: \tilde{X} \to X'$ be a T-equivariant morphism of algebraic varieties and let $U' \subset X'$ be a semi-sectional set in X'. Assume that $\pi|\pi^{-1}(U')$ is an isomorphism. Then $\pi^{-1}(U')$ is a semi-sectional set in $\tilde{X}$.

<u>Proposition 3.8</u>. Let $\tau: \tilde{X} \to X$ be a T-morphism of algebraic varieties and let $U \subset X$ be an open T-invariant subset of X such that $\tau^{-1}(U)$ is a semisectional set in $\tilde{X}$. Then U is semi-sectional in X.

This will finish the proof of Theorem 3.3.

<u>Remark 3.9</u>. A semi-sectional set is sectional if and only if it contains no fixed points. It follows then that if $U \subset X$ is an open T-invariant subset of a normal projective variety X and $U \cap X^T = \emptyset$ and a semi-geometric quotient of U by T exists and is complete, then U is even a sectional set and $U \to U/T$ is a geometric quotient.

<u>Proof of Proposition 3.4</u> By the Chow Lemma we may find a projective variety V with a birational morphism $V \to U/T$. Then $V \times_{U/T} U$ is irreducible (since U is irreducible and $V \to U/T$ is birational). Define $\psi: U' \to V \times_{U/T} U$ as the normalization of $Z = V \times_{U/T} U$.

Then we obtain

$$\begin{array}{ccccc} U' & \xrightarrow{\psi} & Z & \xrightarrow{pr_2} & U \\ \downarrow & & \downarrow{\scriptstyle pr_1} & & \downarrow \\ U'/T & \xrightarrow{\psi'} & V & \longrightarrow & U/T \end{array}$$

Notice that $\Phi = pr_2 \cdot \psi: U' \longrightarrow U$ is proper. Let L be an invertible T-linearized ample sheaf on $V \times_{U/T} U$ such that $Z = (V \times U)^{ss}_{L}$, i.e. Z is composed of semi-stable points with respect to L [GIT]. The existence of such a sheaf follows from the fact that $pr_1: Z \to V$ is a semi-geometric quotient (hence $\phi: Z \to V$ is a categorical quotient and ϕ is affine) and Converse 1.12 p. 41 [GIT]. Let $L' = \psi^*(L)$. Since ψ is an affine finite morphism, we have that $U' = U'^{ss}_{L'}$ so the semi-geometric quotient $U' \to U'/T$ exists. U'/T is normal by Lemma 2.6 because U' is normal and projective (U'/T is quasi-projective by Theorem 1.10 p. 38 [GIT] and $U'/T \to U/T$ is finite since T is reductive)

<u>Proof of Proposition 3.5</u>. Let L' be a very ample invertible T-linearized sheaf on U' such that $U' = U'^{ss}$. Let $\phi_{L'}$ be the T-equivariant embedding of U'

into some projective space $\mathbb{P}^n$ such that $\phi^*_{L'}(\mathcal{O}(1)) = L'$. Obviously $\phi_{L'}(U') \subset \mathbb{P}^{n\,ss}_{\mathcal{O}(1)}$ (we take the T-linearization of $\mathcal{O}(1)$ corresponding to the given linearization of L'). Let

$$\begin{array}{ccc} \eta: X' & \longrightarrow & \hat{X} \subset \mathbb{P}^n \\ & {}_{i}\nwarrow \quad \nearrow_{\phi_{L'}} & \\ & U' & \end{array}$$

be the normalization of the closure $\hat{X}$ of $\phi_{L'}(U')$ in $\mathbb{P}^n$. Then $\eta^*\mathcal{O}(1) = L''$ is a T-linearized very ample sheaf on X' Moreover, $\phi_{L''}(U') \subset \hat{X}^{ss}_{\mathcal{O}(1)}$ so $i(U') \subset X'^{ss}_{L''}$. By Lemma 2.2 we know the set of semi-stable points with respect to a T-linearized ample sheaf is semi-sectional. We also know [GIT theorem 1.10 and remark and the end of page 40] that a semi-geometric quotient of a set of semi-stable points in this case exists, is projective and normal.

Proof of Proposition 3.6. Let $X' = X'_1 \cup \ldots \cup X'_r$ be the decomposition of $(X')^T$ into connected components. Let W be determined by a semi-section (A^+, A^0, A^-) of the set $\{1,\ldots r\}$. We shall define a semi-section (B^+, B^0, B^-) such that U' is the semi-sectional set determined by the semi-section. We have the following diagram:

$$\begin{array}{ccc} U' & \longrightarrow & W \\ \alpha \downarrow & & \downarrow \beta \\ U'/T & \xrightarrow{\ \gamma\ } & W/T \end{array} \qquad \alpha,\ \beta \text{ - semi-geometric quotients}$$

(notice that γ is proper).

Let $i \in A^0$. Then $X'_i \subset W$ and $X'_i \cap U' = \emptyset$ or $X'_i \subset U'$. In fact $\gamma^{-1}(\beta(X'_i)) = \alpha(X'_i \cap U') \cup \gamma^{-1}(\beta(X'_i - U'))$. The set $\beta(X'_i)$ is connected so by Zariski's Main Theorem, we have that $X'_i \cap U' = \emptyset$ or $X'_i - U' = \emptyset$. If $X'_i \cap U' = \emptyset$, then $\gamma^{-1}(\beta(X'_i)) = \alpha((X'_i)^+ \cap U') \cup \alpha((X'_i)^- \cap U')$. Again by Zariski's Main Theorem we infer that $(X'_i)^+ \cap U' = \emptyset$ or $(X'_i)^- \cap U' = \emptyset$ but at least one is nonempty.

Define $B = (B^+, B^0, B^-)$ in the following way $B^+ = A^+ \cup \{i \in A^0;\ X'_i \cap U' = \emptyset$ and $(X'_i)^- \cap U' \neq \emptyset\}$, $B^0 = \{i \in A^0;\ X'_i \subset U'\}$, and $B^- = A^- \cup \{i \in A^0;\ (X'_i)^+ \cap U' \neq \emptyset$ and $X'_i \cap U' = \emptyset\}$. Then (B^+, B^0, B^-) is a semi-section.

Let W' be the semi-sectional set determined by $(B^+, B^0 B^-)$. It is obvious that $U' \subset W' \subset W$. We shall prove that $U' = W'$. Let $x \in W'$. Suppose $\Phi^+(x) \in X'_i$, $\Phi^-(x) \in X'_j$. Then we have $x \in (X'_i)^+ \cap (X'_j)^-$ and we consider the following four cases:

(i) $i \in A^+$ and $j \in A^-$. Then $x \in U'$ because $\alpha^{-1}\gamma^{-1}(\beta(x)) \neq \emptyset$ (and $\alpha^{-1}\gamma^{-1}(\beta(x))$ is exactly one orbit),

(ii) $i, j \in A^0$. Then $i = j$ and $x \in X'_i \subset U'$.

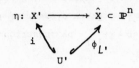

(iii) $i \in A^+$, $j \in A^0$. If $j \in B^0$ then $X'_j \subset U'$ and $x \in U'$, because U' is open (and hence contains $\bar{X}_j$). If $j \in B^+$, then $X'_j \cap U' = \emptyset$ and $U' \cap (X'_j)^- \neq \emptyset$. But $\alpha(U' \cap (X'_j)^-)$ is complete, $\alpha | U' \cap (X'_j)^-$ is a geometric quotient and $\alpha(U' \cap (X'_j)^-)$ is contained in $(X'_j)^-/T$, so $x \in (X'_j) \subset U'$.

(iv) $i \in A^0$ and $j \in A^-$. Then as above we obtain that $x \in U'$.

It follows, that $U' = W'$. □

Proof of Proposition 3.7 and 3.8. Let U' be a ~~semi~~-sectional set in X' corresponding to ~~a~~ ? semi-section (B^+, B^0, B^-). We define a semi-section (C^+, C^0, C^-) in $\tilde{X}$ in the following way: $i \in C^+$ ($i \in C^0$, $i \in C^-$, respectively) if and only if $\pi(\tilde{X}_i)$ is contained in X'_j, where $j \in B^+$ ($j \in B^0$, $j \in B^-$, respectively). Clearly (C^+, C^0, C^-) is a semi-section (notice that we have assumed that $\pi | \pi'(U')$ is an isomorphism).

Similarly, if $\tilde{U}' = \tau^{-1}(U)$ is defined by a semi-section (B^-, B^0, B^+). Then we define a semi-section (C^+, C^0, C^-) for X in the following way: $i \in C^+$ ($i \in C^0$, $i \in C^-$, respectively) if and only if for any connected component $\tilde{X}_j$ contained in $\tau^{-1}(X_i)$, $j \in B^+$ ($j \in B^0$, $j \in B^-$, respectively). Clearly (C^+, C^0, C^-) is a semi-section.

References

[B-B,S] A. Bialynicki-Birula, A. J. Sommese, Quotients of algebraic groups (to appear).

[G.I.T.] D. Mumford, Geometric invariant theory, Ergeb. Math. Bd. 34, Springer Verlag, 1965.

[Se] C. S. Seshadri, Quotient spaces modulo reductive algebraic groups, Ann. of Math. 95(1972), 511-556.

[S] H. Sumihiro, Equivariant completions I, J. Math. Kyoto Univ. 14(1974), 1-28.

A GENERALIZATION OF A THEOREM OF HORROCKS

by

James B. Carrell[1]

and

Andrew John Sommese[2]

In [Hor], G. Horrocks proved that an algebraic action of a solvable group with additive group factors on a connected complete algebraic variety has connected fixed point set. In this article we will show that Horrocks' theorem, and indeed his proof, holds with little change for very general analytic actions. We were lead to consider the generalization in the course of studying $SL(2,\mathbb{C})$ actions on compact Kaehler manifolds [C & S_3].

Theorem. Let $\rho : U \times X \to X$ be a meromorphic action on a compact connected complex space X of a complex solvable linear algebraic groups U with a sequence:

$$0 = U_0 \subseteq U_1 \subseteq \ldots \subseteq U_n \quad n = \dim_{\mathbb{C}} U$$

of subgroups satisfying:

a) U_i is normal in U_{i+1} for $i = 0, \ldots, n-1$,

b) $U_{i+1}/U_i \approx \mathbb{C}$ for $i = 0, \ldots, n-1$.

Then X^U is connected and nonempty.

The fact that X^U is non-empty is just the Borel fixed point theorem; the classical proof carries over with change (cf. (0.2)).

The reason the above theorem carries over is that the Donady family of closures of orbits of a meromorphic $\mathbb{C}$ action on an irreducible compact complex space is compact. This is not immediate because the irreducible components of the Donady space of such an X don't have to be compact.

Horrocks also proves in [Hor] that the pro-finite completions of $\pi_1(X, x)$ and $\pi_1(X^U, x)$ for $x \in X^U$ are isomorphic under the inclusion map. The surjectivity can be proven by Horrocks' arguments. The injectivity is not so clear. In [B-B_2] there is an étale cohomology proof of a slightly weaker result.

§0 Notation and Background Material

By an algebraic solvable group U with factors isomorphic to $\mathbb{C}$ we mean a complex linear algebraic group U with a sequence of algebraic subgroups:

$$0 = U_0 \subseteq U_1 \subseteq \ldots \subseteq U_n = U \quad \text{with} \quad n = \dim_{\mathbb{C}} X$$

satisfying the properties:

---

1. Partially supported by a grant from the Natural Sciences and Engineering Research Council of Canada.

2. Partially supported by N.S.F. Grant MCS-80-03257.

a) U_i is normal in U_{i+1} for $0 \le i \le n-1$

and

b) $U_{i+1}/U_i \approx \mathbb{C}$ for $0 \le i \le n-1$.

Let $\bar{U}$ be a projective manifold with the properties that:

α) U is Zariski open in $\bar{U}$,

and

β) the algebraic action $\rho : U \times U \to U$ by multiplication extends to an algebrai action:

$$\rho : U \times \bar{U} \to \bar{U}.$$

To see that $\bar{U}$ exists note that because U is a linear algebraic group it can be embedded by an algebraic homomorphism ψ of U into $GL(N, \mathbb{C})$ for some N. It is easy to construct a projective manifold $\bar{G}$ for which α) and β) above are true with $(GL(N, \mathbb{C}), \bar{G})$ in place of $(U, \bar{U})$. Let $\bar{U}$ be an equivariant desingularizati of the closure of $\psi(U)$ in $\bar{G}$; the existence of an equivariant desingularization is due to Hironaka.

Let $(U, \bar{U})$ be as above. A meromorphic action $\rho : U \times X \to X$ of U on a compa complex space X is a holomorphic action $\rho : U \times X \to X$ that extends to a meromorphi map $\tilde{\rho} : \bar{U} \times X \to X$.

Note that if $U = \mathbb{C}$ then $\bar{U}$ is $\mathbb{P}^1$.

(0.2) <u>Lemma</u>. <u>Let</u> $\rho : U \times X \to X$ <u>be as above</u>. <u>Then</u> X^U, <u>the fixed point set of</u> U <u>on</u> X <u>is non-empty</u>.

<u>Proof</u>. Let $x \in X$. Let Γ be the graph of $\tilde{\rho}$ in $\bar{U} \times X \times X$. Let

$$A : \Gamma \to U \times X \quad \text{and} \quad B : \Gamma \to X$$

be induced by the projections of $\bar{U} \times X \times X$ onto its first two and last factor respectively.

Note $A^{-1}(U, x)$ is biholomorphic to U. Further $A^{-1}(\bar{U}, x)$ contains a compac analytic set Y in which $A^{-1}(U, x)$ is Zariski open and dense. Therefore $B(Y) = \overline{\rho(U, x)}$ and $\rho(U, x) = B(A^{-1}(U, x))$. From this it is clear that the closures of orbits are analytic sets. Since therefore $\overline{\rho(U, x)} - \rho(U, x)$ is a lower dimensional analytic set of X it can be assumed by descending induction that $\overline{\rho(U, x')} = \rho(U, x'$ for some $x' \in X$. The usual argumentation for the Borel fixed point theorem shows that $\rho(U, x') = x'$.

The following theorem is modelled on a theorem of Fujiki $[F_1]$; we often refer to the family $f : Z \to 2$ of the following theorem as the <u>Fujiki family</u> of closures of orbits. It should be noted that X is not assumed to be a C space. The proof is based on the proof of an analogous result for $\mathbb{C}^*$ actions in [B-B & S]. The proof works for any meromorphic action of a linear algebraic group on a compact complex space X.

(0.3) Theorem. Let $\rho : \mathbb{C} \times X \to X$ be a meromorphic action of $\mathbb{C}$ on an irreducible reduced compact complex space X. Then here is a diagram:

$$\begin{array}{ccc} Z & \xrightarrow{\ \psi\ } & X \\ {\scriptstyle f}\big\downarrow & & \\ \mathcal{Q} & & \end{array}$$

with the following properties:

a) f *is a flat morphism of irreducible compact complex spaces* Z *and* $\mathcal{Q}$;

b) ψ *is a bimeromorphic holomorphic map of* Z *onto* X *such that the restriction of* ψ *to each fibre* $Z_q = f^{-1}(q)$ *is an embedding*;

c) *there is a natural holomorphic action of* $\mathbb{C}$ *on* Z *making* f *and* ψ *equivariant with respect to the trivial* $\mathbb{C}$ *action on* $\mathcal{Q}$ *and* ρ *on* X *respectively*;

d) there is a dense Zanski open set $\mathcal{O} \subseteq \mathcal{Q}$ such that for every $q \in \mathcal{O}$, Z_q is reduced, and $\psi(Z_q)$ is the closure of a $\mathbb{C}$ orbit;

e) every fibre Z_q of f is one dimensional and for fibres $\{Z_q, Z_{q'}\}$ that are reduced, $\psi(Z_q) = \psi(Z_{q'})$ only if $q = q'$; and

f) Z_q is connected for all q.

Proof. Let Γ be the graph in $\mathbb{P}^1 \times X \times X$ of the meromorphic extension $\tilde{\rho} : \mathbb{P}^1 \times X \to X$ of ρ that exists by hypothesis. Let Γ' denote the image of Γ in $X \times X$ under the product projection. Let $a : \Gamma' \to X$ and $b : \Gamma' \to X$ denote the maps of Γ' onto X induced by the projections of $X \times X$ onto its first and second factor respectively. There is a natural action $\gamma : \mathbb{C} \times X \times X \to X \times X$ induced by ρ. It is the product action where:

1) $\mathbb{C}$ acts on the first factor of X by leaving all points fixed,

2) $\mathbb{C}$ acts on the second factor of X by ρ.

Note that:

3) Γ' is invariant under the above action of $\mathbb{C}$.

4) The maps $a : \Gamma' \to X$ and $b : \Gamma' \to X$ are equivariant with respect to this action of $\mathbb{C}$ on Γ' and the actions of $\mathbb{C}$ on X given respectively in 1) and 2) above.

Applying Hironaka's flattening theorem [Hir] to $a : \Gamma' \to X$ and using (3) and (4) we obtain the following:

α) an irreducible compact complex analytic space $\mathcal{X}$ and an irreducible complex analytic subspace G of $\mathcal{X} \times X$,

β) that the holomorphic map $\bar{a} : G \to \mathcal{X}$ and $\bar{b} : G \to X$ (induced by the product projections) which are respectively flat and surjective,

γ) letting $\mathbb{C}$ act trivially on $\mathcal{X}$ and by ρ on X, and by the product action on $\mathcal{X} \times X$, that G is invariant under $\mathbb{C}$ and that $\bar{a}$ and $\bar{b}$ are equivariant.

We let $\mathcal{Q}$ denote the image of $\mathcal{X}$ in the Douady space of $X[D]$ induced by $\bar{a}$. We let $f : Z \to \mathcal{Q}$ denote the flat family over $\mathcal{Q}$. We let ψ denote the induced map to X. Since $\mathcal{X}$ is an irreducible compact complex analytic space, it follows that $\mathcal{Q}$ is irreducible compact complex analytic. This implies that statement a) of the theorem is satisfied.

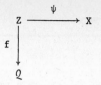

From α), β), and γ) it follows that: with the trivial action of $\mathbb{C}$ on $\mathcal{Q}$, the action ρ on X, and the product action of $\mathbb{C}$ on $\mathcal{Q} \times X$, Z is invariant under $\mathbb{C}$ and f and ψ are $\mathbb{C}$ equivariant. From this we see that c) of the theorem is satisfied. From the definition of a, it follows that there is a dense Zariski open set V of X with the property that for any $v \in V$, $b(a^{-1}(v)) = \{\mathbb{C}x\} \cup \lim_{t\to\infty} \rho(t,x)$ for some $x \in X$. From this and the construction of $\mathcal{Q}$ it follows that there is a dense Zariski open set V' of $\mathcal{Q}$ such that for $q \in V'$ the fibre Z_q of f has an image $\psi(Z_q) = \rho(\mathbb{C}, x)$ for some $x \in X$. By this and the flatness of f, it follows that all fibres of f are one dimensional.

This and the definition of the Donady space imply (e) of the theorem is verified.

There is a dense open set V" of $\mathcal{Q}$ such that the fibres Z_q of f are reduced for $q \in V''$. Let $\mathcal{O} = V'' \cap V'$ where V' is as defined above. The map ψ is surjective since as noted in (β) $\bar{b} : G \to X$ is surjective. From this and (e) it follows that $\psi : f^{-1}(\mathcal{O}) - f^{-1}(\mathcal{O})^{\mathbb{C}} \to X$ is a one to one map of a dense Zariski open set in Z onto a Zariski dense constructible set in X.

Thus ψ is a bimeromorphic holomorphic map of Z onto X which implies (b) of the theorem in view of the definition of the Donady space, is true. This also shows (d).

§1 Proof of Horrocks Theorem

(1.0) Lemma. To prove the assertion:
(H) = (If $\rho : \mathbb{C} \times X \to X$ is a meromorphic action of $\mathbb{C}$ on a compact connected complex space, then $X^{\mathbb{C}}$ is connected), it suffices to prove (H) under the additional assumption that X is irreducible and reduced.

Proof. Clearly X can be assumed to be reduced without loss of generality. Let $\{X_1, \ldots, X_2\}$ be the irreducible components of X. By (H), with the irreducibility assumption, $X_i^{\mathbb{C}}$ is connected for $i = 1, \ldots, n$. To finish the proof of the above lemma it suffices to show that $X_i^{\mathbb{C}} \cap X_j^{\mathbb{C}}$ is non-empty if $X_i \cap X_j$ is non-empty. Assume that $X_i \cap X_j \neq \phi$.

Let $Y_1, \ldots, Y_k$ be the irreducible components of $X_i \cap X_j$. Since $\rho(\mathbb{C}, X_i) \subseteq X$ it follows that $\rho(\mathbb{C}, Y_h) \subseteq Y_h$ for $h = 1, \ldots, k$. By (0.2), $Y_h^{\mathbb{C}}$ is non-empty, i.e. $X_i^{\mathbb{C}} \cap X_j^{\mathbb{C}}$ is non-empty.

(1.1) Lemma. Let $\rho : \mathbb{C} \times X \to X$ be a meromorphic action of $\mathbb{C}$ on a reduced and irreducible compact complex space X. Then $X^{\mathbb{C}}$ is connected.

Proof. Given $x \in X$, the orbit map $\rho(\cdot, x) : \mathbb{C} \to X$ extends to a meromorphic map of $\mathbb{P}^1$ into X. Let $\phi(x)$ denote the image of $\infty \in \mathbb{P}^1$; it can be checked that $\phi(x) \in X^{\mathbb{C}}$.

Since $X^{\mathbb{C}}$ is a compact analytic set, it has finitely many connected components $X_1, X_2, \ldots, X_r$. Let $A_i = \{x \in X \mid \phi(x) \in X_i\}$.

Since X is the disjoint union of $A_1, \ldots, A_r$ it will follow that $r = 1$ if we show that A_i is closed for all i.

Let:

$$\begin{array}{ccc} Z & \xrightarrow{\psi} & X \\ {\scriptstyle f}\downarrow & & \\ \mathcal{Q} & & \end{array}$$

be the Fujiki family of closures of orbits (0.3) of the action ρ. Let $x \in \bar{A}_i$. Let $\langle x_n \rangle$ be a sequence in X with the property that:

$$x_n \longrightarrow x .$$

Let $\langle q_n \rangle \subseteq \mathcal{Q}$ be a sequence satisfying the properties:

(*) $$x_n \in \psi\left(f^{-1}(q_n)\right) \quad \text{and} \quad q_n \to q \in \mathcal{Q}$$

Both properties can be satisfied since ψ is surjective and $\mathcal{Q}$ is compact. Let $Z = \psi\left(f^{-1}(q)\right)$. It is an immediate consequence of the compactness of Z and the continuity of ψ and f that $x \in Z$.

Z is a compact connected one dimensional analytic set that is invariant under the action of $\mathbb{C}$. By (1.0) $Z^{\mathbb{C}}$ is connected if (1.1) is proven for one dimensional X. But this is straightforward. To finish the lemma, it suffices to show that Z meets X_i. To see this note that there is a convergent sub-sequence: $\{y_n\}$ of $\{\phi(x_i)\}$. Here $y_n \to y \in X_i$. An easy argument based on the compactness of Z and the continuity of f and ψ shows that $y_n \to y \in Z$. This completes the proof.

(1.2) Horrocks' Theorem. Let $\rho : U \times X \to X$ be a meromorphic action of a solvable algebraic group with factors isomorphic to $\mathbb{C}$, (cf §0), on a compact connected analytic space X. Then X^U is connected

Proof. Let:

$$0 = U_0 \subseteq U_1 \subseteq \dots \subseteq U_n = U \quad \text{with} \quad n = \dim_{\mathbb{C}} U$$

be a composition series for U satisfying the properties:

a) U_i is normal in U_{i+1},

b) $U_{i+1}/U_i \approx \mathbb{C}$ for $i = 0, \dots, n-1$.

The theorem is true for U_0 trivially. Assume it is true for $0 \le j \le k < n$. By induction we must only show it for $k+1$. Let $X_k = X^{U_k}$. Since U_k is normal in U_{k+1} it follows that ρ descends to an action:

$$\rho_k : U_{k+1}/U_k \times X_k \to X_k .$$

Note that $\mathbb{C} \approx U_{k+1}/U_k$. The reader can check that ρ_k is meromorphic. It follows from (1.1) that $X_k^{\mathbb{C}} = X^{U_{k+1}}$ is connected.

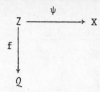

§2 Closing Remarks

Horrocks' theorem has a pleasant application toward the characterization of meromorphic $\mathbb{C}$ actions.

(2.1) Theorem. Let $\rho : \mathbb{C} \times X \to X$ be a holomorphic $\mathbb{C}$ action a connected compact Kaehler manifold X. The action ρ extends to a meromorphic map $\tilde{\rho} : \mathbb{P}^1 \times X \to X$ if and only $X^{\mathbb{C}}$ is non-empty and connected.

Proof. By (0.2) and (1.2) it suffices to show that ρ extends to a meromorphic map $\tilde{\rho} : \mathbb{P}^1_{\mathbb{C}} \times X \to X$ if $X^{\mathbb{C}}$ is non-empty and connected. If $X^{\mathbb{C}}$ is non-empty then by the basic result of $[F_1]$ or [L] the action ρ extends to a holomorphic action:

$$\rho' : G \times X \to X$$

where:

a) G is a connected linear algebraic group,

b) $\mathbb{C}$ is Zariski dense in G,

c) ρ' extends meromorphically to $\bar{G} \times X$ where $\bar{G}$ is any projective manifold in which G embeds as a Zariski open set.

By b) there is the well known consequence that G is commutative and therefore algebraically isomorphic to $(\mathbb{C}^*)^a \times \mathbb{C}^b$ with $b = 0$ or 1. In fact $b \leq 1$ since image of $\mathbb{C}$ in $\mathbb{C}^b$ is Zariski dense and since all subgroups of $\mathbb{C}^b$ are algebraic.

If $b = 0$, then $X^{\mathbb{C}} = X^{(\mathbb{C}^*)^a}$ and since the latter is well known to be disconnected ([B-B$_1$, C&S$_1$, F$_2$]), we get a contradiction. If $b = 1$ and $a > 0$ then $X^{\mathbb{C}} = \left(X^{\mathbb{C}^{*a} \times \langle 0\rangle}\right)^{1\times\mathbb{C}}$. Since $X^{\mathbb{C}^{*a} \times \langle 0\rangle}$ is disconnected and left invariant by the induced meromorphic action of $1 \times \mathbb{C}$, it follows that $X^{\mathbb{C}}$ is disconnected. This contradiction shows that $a = 0$ and $b = 1$.

□

References

[B-B$_1$] A. Bialynicki-Birula, Some theorems on actions of algebraic groups, Annals Math. 98, 480-497 (1973)

[B-B$_2$] ________________, On fixed point schemes of actions of multiplicative and additive groups, Topology 12, 99-103 (1973)

[B-B&S] A. Bialynicki-Birula and A.J. Sommese, Quotients by torus actions, preprint.

[C&S$_1$] J.B. Carrell and A.J. Sommese, Some topological aspects of $\mathbb{C}^*$ actions on compact Kaehler manifolds, Comment. Math. Helvetici 54, 567-582 (1979)

[C&S$_2$] ________________, SL(2, $\mathbb{C}$) actions on compact Kaehler manifolds to appear in Transactions of the Amer. Math. Soc.

[F$_1$] A. Fujiki, On automorphism group of compact Kaehler manifolds, Invent. Math. 44, 225-258 (1978)

[F$_2$] A. Fujiki, Fixed points of the actions on compact Kähler manifolds, Publ. RIMS, Kyoto Univ. 15 (1979), 797-826.

[Hir] H. Hironaka, Flattening theorem in complex-analytic geometry, Amer. Journal of Math. 97, 503-547 (1975)

[Hor] G. Horrocks, Fixed point schemes of additive group actions, Topology 8, 233-242 (1969)

[L] D. Lieberman, Compactness of the Chow scheme: applications to automorphisms and deformations of Kaehler manifolds, Sem. F. Norguet (1977), Lecture Notes in Mathematics 670, New York (1978).

Almost Homogeneous C^* actions on compact complex surfaces

by

James B. Carrell and Andrew John Sommesé

1. Introduction and statement of results.

It is well known (see [C,H,K] and [O,W]) that if X is a compact complex surface which admits a holomorphic C^* action with isolated fixed points, then X is rational. More precisely, there exists a sequence of compact complex spaces X_i with C^* action and equivariant holomorphic maps π_i

(1) $$X = X_0 \xrightarrow{\pi_0} X_1 \xrightarrow{\pi_0} X_2 \longrightarrow \dots \longrightarrow X_n = Z$$

where π_i is the monoidal transformation at a fixed point of X_{i+1} and Z is either $\mathbb{P}^2$, $\mathbb{P}^1 \times \mathbb{P}^1$, or a rational ruled surface $\mathbb{F}_n$ with C^* action having three (in the case of $\mathbb{P}^2$) or four (in the other cases) fixed points.

The purpose of this note is to sharpen (1) by bringing weights into the picture. The weights of a C^* action at a fixed point x are the weights of the representation of C^* in $GL(T_x(X))$ given by $\lambda \mapsto d\lambda_x$. It is well known that when X^{C^*} is isolated, then there exists a unique fixed point s_0 (called the source) at which both weights are positive and a unique fixed point s_∞ (the sink) at which both weights are negative. At any other fixed point, there is one positive and one negative weight. Such fixed points are called hyperbolic.

Definition. A C^* action on a compact complex surface X with isolated fixed points is called almost homogeneous if the weights at the source and sink each have multiplicity two.

By Lemma 3, below, if X has an almost homogeneous C^* action with weights a, a at the source and b, b at the sink, then $a = -b$. We now state the main theorem.

Theorem 1. Suppose X is a compact complex surface having a C^* action with isolated fixed point set X^{C^*}. Let a, b be the weights at the source of X and let $c = \text{g.c.d.}(a,b)$. Then there exists a compact complex surface $\tilde{X}$ with almost homogeneous C^* action having weights c at the source and $-c$ at the sink and a relatively minimal complex surface Z with homogeneous C^* action with weights c at the source and $-c$ at the sink and a diagram of surjective holomorphic equivariant maps

$$\begin{array}{ccc} & \tilde{X} & \\ \swarrow & & \searrow \\ X & & Z \end{array}$$

Moreover, Z is either $\mathbb{P}^1 \times \mathbb{P}^1$ or $\mathbb{F}_2$ and the weights of the C^* action on Z are (a,a), $(-a,a)$, $(-a,a)$, and $(-a,-a)$.

Remark. If one blows up $\tilde{X}$ at the source and the sink, then one obtains Y with C^* action having source and sink $\mathbb{P}^1$. Thus one is in the situation considered in [O,W]. An interesting point is that, by Theorem 1, the graph associated to Y in [O,W] has either one or two arms (corresponding to the singular chains in Y). Hency by [O,W], Y admits a $C^* \times C^*$ action. Hence by an easy argument, so does X (for any $C^* \times C^*$ action is equivariant with respect to blowing down).

To prove Theorem 1, we will use

Theorem 2. Let Y be as in the remark. Then there exists a holomorphic equivariant projection $\psi: Y \to \mathbb{P}^1$ (with the trivial C^* action) which is an isomorphism on the source and sink of Y. ψ has at most two singular fibres. These are the singular chains in Y.

For any orbit $\mathcal{O} = C^* \cdot x$, let $\mathcal{O}_0 = \lim_{\lambda \to 0} \lambda \cdot x$ and $\mathcal{O}_\infty = \lim_{\lambda \to \infty} \lambda \cdot x$. These limits exist since the finiteness of X^{C^*} implies that X is projective. By a <u>singular chain</u> in X we mean a sequence of orbits $\mathcal{O}_1, \ldots, \mathcal{O}_k$ so that $(\mathcal{O}_1)_0 = s_0$ (the source), $(\mathcal{O}_k)_\infty = s_\infty$ (the sink) and $(\mathcal{O}_i)_\infty = (\mathcal{O}_{i+1})_0$ for $1 \le i \le k-1$.

Theorem 3. If X^{C^*} is finite, then there exist at most two singular chains in X and every hyperbolic fixed point lies on exactly one of these chains. The set of varieties $\overline{\mathcal{O}}$ where $\mathcal{O}$ is an orbit in X so that $(\mathcal{O}_i)_0$ is a hyperbolic fixed point form a homology basis of $H_2(X,\mathbb{Z})$. In particular $b_2(X)$ is the number of hyperbolic fixed points in X. Finally there are two singular chains if and only if the minimal model Z of X is $\mathbb{P}^1 \times \mathbb{P}^1$.

2. <u>Some Lemmas on weights.</u>

Lemma 1. The weights of any C^* action on $\mathbb{P}^2$ with isolated fixed points are of the form

(3) $$(a,b)\ ,\ (-a,b-a)\ ,\ (a-b,-b)$$

for some distinct positive integers a and b. The weights of a C^* action on a rational ruled surface $\mathbb{F}_n$ with isolated fixed points are

(4) $$(a,b)\ ,\ (a,-b)\ ,\ (-a,b-na)\ ,\ (-a,na-b)$$

where a and b are distinct positive integers so that $na \neq b$. The weights of any C^* action on $\mathbb{P}^1 \times \mathbb{P}^1$ with isolated fixed points are

(5) $$(a,b)\ ,\ (a,-b)\ ,\ (-a,b)\ ,\ (-a,-b)$$

where a, b are any positive integers.

The proof of this lemma is in [C,H,K] .

Lemma 2. Let $\pi : \tilde{X} \to X$ denote the monoidal transform of X at a fixed point $x \in X$. Then the action on X lifts to $\tilde{X}$ so that π is equivariant. If a, b denote the weights at x and $a \neq b$, then $\tilde{X}$ has two fixed points on $\pi^{-1}(x)$ with weights

$$(a,b-a) \text{ and } (b,a-b)$$

respectively. If $a = b$, then $\pi^{-1}(x)$ is a component of $\tilde{X}^{C^*}$ with weights (a,0) .

Proof. Let (u,v) denote equivariant local coordinates on a neighborhood W of x so that $x = (0,0)$. Thus $\lambda\cdot(u,v)=(\lambda^a u,\lambda^b v)$. By definition, a neighborhood V of $\pi^{-1}(x)$ consists of all points $(u,v,[x,y]) \in W \times \mathbb{P}^1$ such that $uy = vx$. C* acts on V by $\lambda\cdot(u,v,[x,y]) = (\lambda^a u,\lambda^b v,[\lambda^a x,\lambda^b y])$. This action extends to $\tilde{X}$ making π equivariant. If $a \neq b$, then V^{C^*} consists of (0,0,[1,0]) and (0,0,[0,1]) . Since local equivariant coordinates near (0,0,[1,0]) are (u,y/x) , it follows that the weights at (0,0,[1,0]) are (a,b-a) . Similarly, the weights at (0,0,[0,1]) are (b,a-b) . If $a = b$, then $\pi^{-1}(x)$ is a component of V^{C^*} with weights (a,0) .

Lemma 3. Let X have weights (a,b) at s_0 and weights (c,d) at s_∞ . Let $\tilde{X}$ be obtained from X by blowing up sources and sinks until the weights at the source $\tilde{s}_0$ of $\tilde{X}$ are (q,q) and at the sink $\tilde{s}_\infty$ are (r,r) . Then $q = \text{g.c.d.}(a,b)$, $r = \text{g.c.d.}(c,d)$, and $r = -q$.

Proof. That $\tilde{X}$ exists and that $q = \text{g.c.d.}(a,b)$ and $r = \text{g.c.d.}(c,d)$ follow from Lemma 2. To finish the proof we will show $q = -r$. Let $\mathcal{O}$ be an orbit in $\tilde{X}$ so that $\mathcal{O}_0 = \tilde{s}_0$ and $\mathcal{O}_\infty = \tilde{s}_\infty$. Let $y \in \mathcal{O}$ and consider the isotropy group of y . Since $\lambda \in C^*$ acts with unique weight q on $T_{\tilde{s}_0}(\tilde{X})$, it follows that the isotropy group of y consists of the q-th roots of unity. But as $\mathcal{O}_\infty = \tilde{s}_\infty$, the q-th roots of unity and r-th roots of unity must coincide. Hence $q = -r$.

Lemma 4. Let $\mathcal{O}_1, \mathcal{O}_2, \ldots, \mathcal{O}_k$ be a singular chain in X . Then if the weight $a > 0$ occurs at $(\mathcal{O}_i)_\infty$, $1 \leq i < k$, the weight $-a$ occurs at $(\mathcal{O}_{i+1})_0$.

Proof. To see this statement, argue on the isotropy groups as in Lemma 3.

§3. Proofs of the theorems.

We will first prove Theorem 2. Let Y be obtained from blowing up the source and sink of $\tilde{X}$. Thus the source F^+ and sink F^- of Y are both $\mathbb{P}^1$'s . Let $\mathcal{O}_1$ and $\mathcal{O}_2$ be any pair of disjoint orbits from the source to the sink. Then $\overline{\mathcal{O}}_1$ and $\overline{\mathcal{O}}_2$ are both P^1's that have the same line bundle L . Hence we may take disjoint sections of L to obtain an equivariant holomorphic map ψ to $\mathbb{P}^1$. The bundle L has degree one on each fibre. But $\mathcal{O}(1)$ is spanned by exactly two sections so ψ is biholomorphic on F^+ and F^- . At most two fibres of ψ fail to be nonsingular. These fibres are precisely the singular chains in $\tilde{X}$ and we know there are at most two singular chains in $\tilde{X}$ since the minimal model of $\tilde{X}$ has at most two singular chains by Lemma 4. This completes the proof of Theorem 2.

We will now prove Theorem 1. By Lemmas 2 and 3, we may as well assume $\tilde{X} = X$. We know that $\tilde{X}$ can be equivariantly blown down to a relatively minimal surface. In fact, if C is an exceptional curve of the first kind in X, C is C^* invariant since $C \cdot C < 0$. Hence exceptional curves of the first kind are closures of orbits and, by Lemma 2, lie on a singular chain (since if C can be blown down, C cannot meet both the source and the sink). Thus we want to show that on any singular chain $O_1 + \ldots + O_k$, we can make the chain relatively minimal by blowing down certain O_i with $1 < i < k$. If neither $\overline{O}_1$ nor $\overline{O}_k$ can be blown down, and if the chain is not relatively minimal, then we may blow down a suitable $\overline{O}_i$ with $1 < i < k$ without affecting almost homogeneity of the action. Thus we may continue in this manner until chain is relatively minimal or either $\overline{O}_1$ or $\overline{O}_k$ can be blown down.

Now suppose $\overline{O}_1$ can be blown down. Then it follows that the weights at $(O_1)_\infty$ are $2a, -a$ by Lemma 2. Now blow up s_0 and s_∞ to obtain Y as in Theorem 2. Let

$$F = \lambda_1 A_1 + \lambda_2 A_2 + \ldots + \lambda_k A_k$$

be the singular fibre in Y associated to $O_1 + \ldots + O_k$, where A_i is the proper transform of O_i. Each A_i is the closure of an orbit in Y and $\lambda_i (>0)$ is the order of the isotropy group of A_i. Thus for example $\lambda_1 = \lambda_k = c$ and $\lambda_2 = 2c$. Now by flatness of $\psi : Y \to \mathbb{P}^1$, any two fibres of ψ are homologous. Since A_i misses the nonsingular fibres, $F \cdot A_i = 0$ for $1 \le i \le k$. Thus if $p_i = A_i \cdot A_i$,

$$\text{(6)} \qquad \begin{aligned} &\lambda_1 p_1 + \lambda_2 = 0 \\ &\lambda_{i-1} + p_i \lambda_i + \lambda_{i+1} = 0 \quad \text{for} \quad 1 < i < k \\ &\lambda_{k-1} + p_k \lambda_k = 0 . \end{aligned}$$

(6) follows since $A_i \cdot A_{i+1} = 1$ while $A_i A_j = 0$ if $|i-j| > 1$. We must show that $p_i = -1$ for some i with $1 < i < k$. Suppose all $p_i \le -2$ for $1 < i < k$. Then

$$\lambda_i \le \frac{1}{2}(\lambda_{i-1} + \lambda_{i+1}) , \quad 1 < i < k .$$

Summing gives

$$\sum_{i=2}^{k-1} \lambda_i \le \frac{1}{2} \sum_{i=1}^{k-2} \lambda_i + \frac{1}{2} \sum_{i=3}^{k} \lambda_i$$

so

$$\lambda_2 + \lambda_{k-1} \le \frac{1}{2} (\lambda_1 + \lambda_2 + \lambda_{k-1} + \lambda_k)$$

nd thus

$$\lambda_2 + \lambda_{k-1} \le \lambda_1 + \lambda_k$$

ut $\lambda_2 = 2a$, $\lambda_{k-1} > 0$ and $\lambda_1 = \lambda_k = a$. This is impossible, so $p_i = -1$ for some with $1 < i < k$. After blowing down A_i , blow down F^+ and F^- and repeat the rgument on the new homogeneous action X' . Clearly the process eventually leads o a relatively minimal surface Z with almost homogeneous C* action as asserted. hat Z is either $\mathbb{P}^1 \times \mathbb{P}^1$ or $\mathbb{F}_2$ with weights as asserted follows from (4) and 5) of Lemma 1. This completes the proof of Theorem 1.

The proof of Theorem 3 follows from [C,S] and remarks above.

References

C,H,K] J. Carrell, A. Howard, and C. Kosniowski. Holomorphic vector fields on complex surfaces, Math. Ann. 204, 73-81 (1973)

C,S] J. Carrell and A.J. Sommese. Some topological aspects of C* actions on compact Kaehler manifolds, Comment. Math. Helvetic. 54, 567-582 (1979)

O,W] P. Orlik and P. Wagreich. Algebraic surfaces with k* action, Acta Math. 138, 43-81 (1977)

WEIGHTED PROJECTIVE VARIETIES

by

Igor Dolgachev

Contents

0. Introduction
1. Weighted projective space
   1.1. Notations
   1.2. Interpretations
   1.3. The first properties
   1.4. Cohomology of $\mathcal{O}_{\mathbb{P}}(n)$
   1.5. Pathologies
2. Bott's theorem
   2.1. The sheaves $\bar{\Omega}^i_{\mathbb{P}}(n)$
   2.2. Justifications
   2.3. Cohomology of $\bar{\Omega}^i_{\mathbb{P}}(n)$
3. Weighted complete intersections
   3.1. Quasicones
   3.2. Complete intersections
   3.3. The dualizing sheaf
   3.4. The Poincare series
   3.5. Examples
4. The Hodge structure on cohomology of weighted hypersurfaces.
   4.1. A resolution of $\tilde{\Omega}^i_X$
   4.2. The Griffiths theorem
   4.3. Explicit calculation
   4.4. Examples and supplements

0. Introduction.

In this paper I discuss the technique of weighted homogeneous coordinates which has appeared in works of various geometers a few years ago and it seems has been appreciated and armed by many people. In many cases this technique allows one to present a nonsingular algebraic variety as a hypersurface in a certain space (a weighted projective space) and deal with it as it would be a nonsingular hypersurface in the projective space. A generalization of this approach is the technique of polyhedral projective spaces for which we refer to [5, 6, 15].

Section 1 deals with weighted projective spaces, the spectrums of graded polynomial rings. Most of the results from this section can be found in [7].

In section 2 we generalize the Bott theorem on the cohomology of twisted sheaves of differentials to the case of weighted projective spaces. Another proof of the same result can be found in [23] and a similar result for torical spaces is discussed in [5].

In section 3 we introduce the notion of a quasismooth subvariety of a weighted projective space. For this we define the affine quasicone over a subvariety and require that this quasicone is smooth outside its vertex. We show that quasismooth weighted complete intersections have many properties of ordinary smooth complete intersections in a projective space. The work of Mori [19] contains a similar result but under more restrictive conditions. Rather surprisingly not everything goes the same as for smooth complete intersections. For example, recent examples of Catanese and Todorov show that the local Torelli theorem fails for some quasismooth weighted complete intersections (see [4, 24]).

In section 4 we generalize to the weighted case the results concerning the Hodge structure of a smooth projective hypersurfaces. Our proof is an algebraic version of one of Steenbrink [23] and can be applied to the calculation of the De Rham cohomology of any such hypersurface over a field of characteristic zero. The present paper is partially based on my talks at a seminar on the Hodge-Deligne theory at Moscow State University in 1975/76. It is a pleasure to thank all of its participants for their attention and criticism.

1. Weighted projective space.

1.1. Notations

$Q = \{q_0, q_1, \ldots, q_r\}$, - a finite set of positive integers;

$$|Q| = q_0 + \ldots + q_r ;$$

$S(Q)$ - the polynomial algebra $k[T_0, \ldots, T_r]$ over a field k, graded by the condition $\deg(T_i) = q_i$, $i = 0, \ldots, r$; $\mathbb{P}(Q) = \mathrm{Proj}(S(Q))$ - weighted projective space of type Q. $U = \mathbb{A}^{r+1} - \{0\} = \mathrm{Spec}(S(Q)) - \{(T_0, \ldots, T_r)\}$; $m = (T_0, \ldots, T_r)$.

Abbreviations:

$\mathbb{P}^r = \mathbb{P}(1, \ldots, 1)$, $S = S(Q)$, $\mathbb{P} = \mathbb{P}(Q)$.

We suppose in the sequel that the characteristic p of k is prime to all q_i, though many results are valid without this assumption. We also assume that $(q_0, \ldots, q_r) = 1$.

The last assumption is not essential in virtue of the following:

Lemma. Let $Q' = \{aq_0, \ldots, aq_r\}$. Then $\mathbb{P}(Q) \simeq \mathbb{P}(Q')$.

Really, $S(Q')_m = S(Q)_{am}$ and hence in the standard notations of [12] we have $S(Q') = S(Q)^{(a)}$. Applying ([12], 2.4.7) we obtain a canonical isomorphism

$$\mathbb{P}(Q) = \mathrm{Proj}(S(Q)) \simeq \mathrm{Proj}(S(Q)^{(a)}) = \mathbb{P}(Q').$$

We refer to 1.3 for more general results.

For any graded module M over a graded commutative ring A we denote by $M(n)$ the graded A-module obtained by shifting the graduation $M(n)_k = M_{n+k}$.

By $\tilde{M}$ we denote the $\mathcal{O}_{\mathrm{Proj}(A)}$-Module, associated with M. Recall ([12]; 2.5.2) that for any $f \in A_d$

$$\Gamma(D_+(f), \tilde{M}) = M_{(f)} = \{\frac{m}{f^k} : m \in M_{kd}\} ,$$

where open sets $D_+(f) = \mathrm{Spec}(A_{(f)})$ form a base of open sets in $\mathrm{Proj}(A)$.

1.2. Interpretations.

1.2.1. It is well known that a $\mathbb{Z}$-graduation of a commutative ring is equivalent to an action of a 1-dimensional algebraic torus G_m on its spectrum. In our case G_m acts on $\mathbb{A}^{r+1} = \mathrm{Spec}(S(Q))$ as follows

$$S \longrightarrow S \otimes k[X, X^{-1}]$$

$$T_i \longrightarrow T_i \otimes X^{q_i}, \quad i = 0, \ldots, r$$

where $G_m = \mathrm{Spec}(k[X, X^{-1}])$.

The corresponding action on points with the value in a field $k' \supset k$ is given by the formulas

$$k'^* \times k'^{r+1} \to k'^{r+1}$$

$$(t, (a_0, \ldots, a_r)) \to (a_0 t^{q_0}, \ldots, a_r t^{q_r})$$

The open set $U = \mathbb{A}^{r+1} - \{0\}$ is invariant with respect to this action and the universal geometric quotient U/G_m exists and coincides with $\mathbb{P}(Q)$.

If $k = \mathbb{C}$ is the field of complex numbers then the analytic space $\mathbb{P}^{an}$ associated to $\mathbb{P}(Q)$ is a complex analytic quotient space $\mathbb{C}^{r+1} - \{0\}/\mathbb{C}^*$ where $\mathbb{C}^*$ acts on $\mathbb{C}^{r+1}$ by the formulas

$$(t, (z_0, \ldots, z_r)) \to (z_0 t^{q_0}, \ldots, z_r t^{q_r}) .$$

In view of this interpretation the space $\mathbb{P}(Q')$ from the lemma in 1.1 corresponds to a noneffective action of G_m.

1.2.2. For any positive integer q we denote by μ_q the finite group scheme of q-roots of unity. This is a closed subgroup of G_m with the coordinate ring $k[X]/(X^q - 1)$.

Consider the action of the group scheme $\mu_Q = \mu_{q_0} \times \ldots \times \mu_{q_r}$ on $\mathbb{P}^r$ which is induced by the action μ_Q on S

$$T_i \to T_i \otimes \overline{X}_i ,$$

where $\overline{X}_i \equiv X \bmod (X^{q_i} - 1)$ in the coordinate ring of μ_{q_i}.

The homomorphism of rings $S(Q) \to S$, $T_i \to T_i^{q_i}$ yields the isomorphism

$S(Q) \simeq S^{\mu}Q$. It is easy to see that the corresponding morphism of projective spectrums is well defined and gives an isomorphism

$$\mathbb{P}(Q) \simeq \mathrm{Proj}(S^{\mu}Q) \simeq \mathbb{P}^r/\mu_Q .$$

In case $k = \mathbb{C}$

$$\mathbb{P}(Q)^{an} = \mathbb{P}^r(\mathbb{C})/\mu_Q(\mathbb{C})$$

where $\mu_Q(\mathbb{C})$ acts by the formulas

$$(g,(z_0,\ldots,z_r)) \to (z_0g_0,\ldots,z_rg_r)$$

$$g = (g_0,\ldots,g_r), \quad g_i = \exp(2\pi i b_i/q_i), \quad 0 \le b_i < q_i .$$

1.2.3. The previous interpretation easily gives, for instance, that for $Q = \{1,1,\ldots,1,n\}$ the weighted projective space $\mathbb{P}(Q)$ equals the projective cone over the Veronese variety $v_n(\mathbb{P}^{r-1})$.

For example, $\mathbb{P}(1,1,n)$, $n \neq 1$ is obtained by the blowing down the exceptional section of the ruled surface $\mathbb{F}_n$ (when $n=2$ it is an ordinary quadratic cone).

1.2.4. For $Q = \{1,q_1,\ldots,q_r\}$ the spaces $\mathbb{P}(Q)$ are compactifications of the affine space $\mathbb{A}^r$. Indeed, the open set $D_+(T_0)$ is isomorphic to the spectrum of the polynomial ring $k\left[\frac{T_1}{T_0},\ldots,\frac{T_r}{T_0}\right]$. Its complement coincides with the weighted projective space $\mathbb{P}(q_1,\ldots,q_r)$.

1.2.5. Weighted projective spaces are complete toric spaces. More precisely, $\mathbb{P}(q_0,\ldots,q_r)$ is isomorphic to the polyhedral space $\mathbb{P}_\Delta$ of [6], where $\Delta = \{(x_0,\ldots,x_r) \in \mathbb{R}^{r+1} : \Sigma q_i x_i = q_0 \cdots q_r\}$.

1.3. <u>The first properties</u>

1.3.1. For different Q and Q' the corresponding spaces $\mathbb{P}(Q)$ and $\mathbb{P}(Q')$ can be isomorphic.

Let

$$d_i = (q_0,\ldots,q_{i-1},q_{i+1},\ldots,q_r)$$

$$a_i = \ell.\mathrm{c.m.}(d_0,\ldots,d_{i-1},d_{i+1},\ldots,d_r)$$

$$a = \ell.\mathrm{c.m.}(d_0,\ldots,d_r).$$

Note that $a_i | q_i$, $(a_i,d_i) = 1$ and $a_i d_i = a$.

<u>Proposition</u>. (Delorme [7].) Let $Q' = \{q_0/a,\ldots,q_r/a_r\}$. Then there exists a natural isomorphism $\mathbb{P}(Q) \simeq \mathbb{P}(Q')$.

For the proof we consider the graded subring $S' = \bigoplus_{n=0}^{\infty} S(Q)_{an}$ of $S(Q)$ and note that $S' \simeq k[X_0,\ldots,X_r]$, where $X_i = T_i^{d_i}$ is of degree aq_i/a_i. But then

$S(Q') \simeq S'^{(a)}$ and hence $\mathrm{Proj}(S(Q')) \simeq \mathrm{Proj}(S')$ ([12],2.4.7). Now there exists also an isomorphism $\mathrm{Proj}(S') \simeq \mathrm{Proj}(S(Q)^{(a)}) \simeq \mathrm{Proj}(S(Q))$.

Corollary. Each $\mathbb{P}(Q) \simeq \mathbb{P}(Q')$, where $(q'_0,\ldots,q'_{i-1},q'_{i+1},\ldots,q'_r) = 1$ for $i = 0,\ldots,r$.

Corollary. Assume that $q_i = a_i$ for $i = 0,\ldots,r$. Then $\mathbb{P}(Q) \simeq \mathbb{P}^r$.

For example, it is so if all numbers $\ell.c.m.(q_0,\ldots,q_r)/q_i$ are coprime. In this case the previous fact was independently discovered by M. Reid.

Note that in case $r = 1$ we can use the previous corollary and obtain that $\mathbb{P}(q_0,q_1) \simeq \mathbb{P}^1$ for any q_0,q_1. This fact however follows also from interpretation 1.2.2.

1.3.2. Remarks. 1. There is a certain difference between the identifications of the proposition and of the lemma in 1.1. In terms of 3.5, the spaces $\mathbb{P}(Q)$ and $\mathbb{P}(Q')$ from the proposition are not projectively isomorphic.

2. It can be shown that the isomorphism $\mathbb{P}(Q) \simeq \mathbb{P}(Q')$ of 1.3.1 induces an isomorphism of sheaves $\mathcal{O}_{\mathbb{P}}(n) \simeq \mathcal{O}_{\mathbb{P}'}((n - \sum_{i=0}^{r} b_i(n)q_i)/a)$, where $b_i(n)$ are uniquely determined by the property

$$n = b_i(n)q_i + c_i(n)d_i, \quad 0 \le b_i < d_i .$$

1.3.2. Let G be a finite group of linear automorphisms of a finite-dimensional vector space V over a field k. An element $g \in G$ is called a pseudoreflection if there exists an element $e_g \in V$ and $f_g \in V^*$ such that

$$g(x) = x + f_g(x)e_g \quad \text{for every } x \in V .$$

Lemma. ([3], ch.V, §5, th.4.) Let B be the symmetric algebra of V and $A = B^G$, the subalgebra of G-invariant elements. Assume that $\#G$ is invertible in k. Then the following assertions are equivalent:

(i) G is generated by pseudoreflections;

(ii) A is a graded polynomial k-algebra.

Example. μ_Q acts on S as a group generated by pseudoreflections. These pseudoreflections act by the formula

$$T_i \to T_i \otimes \overline{X}_i ,$$
$$T_j \to T_j , \quad j \neq i , \quad i = 0,\ldots,n .$$

1.3.3. Proposition

(i) $\mathbb{P}(Q)$ is a normal irreducible projective algebraic variety;

ii) all singularities of $\mathbb{P}(Q)$ are cyclic quotients singularities (in particular, $\mathbb{P}(Q)$ is a V-variety);

iii) a nonsingular $\mathbb{P}(Q)$ is isomorphic to $\mathbb{P}^r$.

For the proof of property (i) we remark that this property is preserved under an action of a finite group and use interpretation 1.2.2. To see (ii), we use interpretation 1.2.1. Let $\mathbb{P} = \bigcup_{i=0}^{r} U_i$ be the canonical covering of $\mathbb{P}$, where $U_i = D_+(T_i)$. Consider the closed subvariety $V_i = \mathrm{Spec}(S/(T_i - 1))$ of $\mathbb{A}^{r+1} = \mathrm{Spec}(S)$. The action of G_m on $\mathbb{A}^{r+1}$ induces the action of μ_{q_i} on V_i which, after identifying V_i with $\mathrm{Spec}(k[T_0,\dots,T_{i-1},T_{i+1},\dots,T_r])$, can be given by the formulas

$$T_j \to T_j \otimes \overline{X}_i^{q_j}, \quad j = 0,\dots,i-1,i+1,\dots,r$$

where notations as in 1.2.2.

It is easy to see that $U_i \simeq V_i/\mu_{q_i}$ and, since $V_i \simeq \mathbb{A}^r$, we have property (ii) of $\mathbb{P}(Q)$.

For the proof of (iii) we use 1.3.1 and the previous construction. By 1.3.1 we may assume that $(q_0,\dots,q_{i-1},q_{i+1},\dots,q_r) = 1$. Then it is easy to see that the action of μ_{q_i} on V_i is generated by pseudoreflections only in the case $q_i = 1$. It remains to apply 1.3.2.

1.4. Cohomology of $\mathcal{O}_{\mathbb{P}}(n)$.

1.4.1. Recall that $\mathcal{O}_{\mathbb{P}}(n)$ denotes an $\mathcal{O}_{\mathbb{P}}$-Module associated to the graded $S(Q)$-module $S(Q)(n)$. For any homogeneous $f \in S(Q)$ we have a natural homomorphism $S(Q)_n \to S(Q)(n)_{(f)}$ $(a \to a/1)$ which defines a natural homomorphism $\alpha_n : S(Q)_n \to H^0(\mathbb{P}(Q),\mathcal{O}_{\mathbb{P}}(n))$ (the Serre homomorphism).

Theorem.

i) $\alpha_n : S_n \to H^0(\mathbb{P}, \mathcal{O}_{\mathbb{P}}(n))$ is bijective for any $n \in \mathbb{Z}$;

ii) $H^i(\mathbb{P},\mathcal{O}_{\mathbb{P}}(n)) = 0$ for $i \neq 0,r$, $n \in \mathbb{Z}$;

iii) $H^r(\mathbb{P},\mathcal{O}_{\mathbb{P}}(n)) \simeq S_{-n-|Q|}$.

Proof. According to general properties of projective spectrums we can identify $U = \mathrm{Spec}(S) - \{m\}$ with the affine spectrum of the graded $\mathcal{O}_{\mathbb{P}}$-Algebra $\bigoplus_{n\in\mathbb{Z}} \mathcal{O}_{\mathbb{P}}(n)$ ([12], 8.3). The corresponding projection $p: U \to \mathbb{P}$ coincides with the quotient morphism $U \to U/G_m$ from 1.2.1. Since p is an affine morphism we have

$$H^i(U,\mathcal{O}_U) \simeq H^i(\mathbb{P}, p_*(\mathcal{O}_U)) \simeq H^i(\mathbb{P}, \bigoplus_{n\in\mathbb{Z}} \mathcal{O}_{\mathbb{P}}(n)) \simeq \bigoplus_{n\in\mathbb{Z}} H^i(\mathbb{P},\mathcal{O}_{\mathbb{P}}(n)).$$

Now we use the local cohomology theory ([13]). We have an exact sequence

$$0 \to H^0_{\{m\}}(S) \to S \to H^0(U,\mathcal{O}_U) \to H^1_{\{m\}}(S) \to 0$$

and isomorphisms

$$H^i_{\{m\}}(S) \simeq H^{i-1}(U,\mathcal{O}_U), \quad i>1.$$

It is easy to see that the homomorphism $S \to H^0(U,\mathcal{O}_U)$ induces on each S_n the Serre homomorphism $\alpha_n : S_n \to H^0(\mathbb{P},\mathcal{O}_{\mathbb{P}}(n))$. Since S is a Cohen-Macaulay ring, we have

$$H^i_{\{m\}}(S) = 0, \quad i \neq r+1.$$

This proves assertions (i), (ii) of the theorem.

For the proof of (iii) we have to use the explicit calculation of $H^{r+1}_{\{m\}}(S)$. We have

$$H^{r+1}_{\{m\}}(S) = \varinjlim_m \operatorname{Ext}^{r+1}(S/(T_0,\dots,T_r)^m,S) =$$
$$= \varinjlim_m \operatorname{Ext}^{r+1}(S/(\underline{T}^m),S)$$

where $(\underline{T}^m) = (T_0^m,\dots,T_r^m)$.

Let V be a free S-module of rang $r+1$ with the basis $(e_0,\dots,e_r)$. Grade V by the condition $\deg(e_i) = q_i m$ and consider the induced gradation on its exterior powers $\overset{p}{\Lambda}(V)$ (where $\deg(e_{i_1}\wedge\dots\wedge e_{i_p}) = m(q_{i_1}+\dots+q_{i_p})$). The Koszul complex for $(\underline{T}^m)$:

$$S \leftarrow \overset{1}{\Lambda}(V) \leftarrow \overset{2}{\Lambda}(V) \leftarrow \dots\dots \leftarrow \overset{r+1}{\Lambda}(V) \leftarrow 0$$

$$e_{i_1}\wedge\dots\wedge e_{i_p} \to \sum_k (-1)^k T^m_{i_k} e_{i_1}\wedge\dots\wedge\hat{e}_{i_k}\wedge\dots\wedge e_{i_p}$$

defines a resolution of graded S-modules for $S/(\underline{T}^m)$ and hence we have an isomorphism of graded S-modules

$$\operatorname{Ext}^{r+1}(S/(\underline{T}^m),S) \simeq \operatorname{Hom}(\overset{r+1}{\Lambda}(V),S)/\operatorname{Im}(\operatorname{Hom}(\overset{r}{\Lambda}(V),S) \simeq$$
$$\simeq (S/(\underline{T}^m))(-m|Q|)$$

Put

$$I_m = (S/(\underline{T}^m))(-m|Q|),$$

then

$$H^{r+1}_{\{m\}}(S) = \varinjlim_m I_m,$$

where the inductive system is described as follows.

Let $t^m_{a_0,\dots,a_r}$ be the image of $T_0^{m-a_0}\dots T_r^{m-a_r}$ in I_m. It is clear that for $0 < a_i \le m$ $t^m_{a_0,\dots,a_r}$ form a basis of I_m. In this notation the transition map

$$u_{m,m+s} : I_m \to I_{m+s}$$

is multiplication by $T_0^s\dots T_r^s$ and

$$u_{m,m+s}(t^m_{a_0,\dots,a_r}) = t^{m+s}_{a_0+s,\dots,a_r+s}.$$

Let $\bar{e}_{a_0,\dots a_r}$ be the image of $e^m_{a_0,\dots,a_r}$ in $\varinjlim_m I_m$. Module $H^{r+1}_{\{m\}}(S)$ is a graded S-module and elements $\overline{e}_{a_0,\dots,a_r}$ form its homogeneous basis. We have

$$\deg(\overline{e}_{a_0,\dots,a_r}) = \deg(e^m_{a_0,\dots,a_r}) = (m-a_0)+\dots+(m-a_r)-m|Q| = \sum_{i=0}^{r} a_i q_i$$

Thus we obtain that $\overline{e}_{a_0,\dots,a_r}$ with

$$n = -\sum_{i=0}^{r} a_i q_i \quad (a_i > 0)$$

generate $H^{r+1}_{\{m\}}(S)_n$ as a k-space. Since

$$\dim_k S_{-n-Q} = \#\{(b_0,\dots,b_r) \in \mathbb{N}^{r+1} : -n-|Q| = \sum_{i=0}^{r} b_i q_i\} =$$

$$= \#\{(a_0,\dots,a_r) \in \mathbb{N}^{r+1}_{+} : -n = \sum_{i=0}^{r} a_i q_i\} ,$$

we have

$$H^{r+1}_{\{m\}}(S)_n \simeq S_{-n-|Q|}.$$

It remains to notice that

$$H^{r+1}_{\{m\}}(S)_n \simeq H^r(\mathbb{P}, \mathcal{O}_{\mathbb{P}}(n)) .$$

1.4.2. Let integers a_n be determined by the identity

$$P_S(t) = \sum_{n=0}^{\infty} a_n t^n = \prod_{i=0}^{r} (1-t^{q_i})^{-1} .$$

Then as a corollary of the previous theorem we have

$$\dim_k H^i(\mathbb{P}, \mathcal{O}_{\mathbb{P}}(n)) = \begin{cases} a_n & i = 0 \\ 0 & 1 \neq 0, r \\ a_{-n-|Q|} , & i = r \end{cases}$$

In fact, $P_S(t)$ is the Poincare series of the graded algebra $S(Q)$ (see 3.4) and $a_n = \dim_k S(Q)_n$.

1.5. Pathologies

If $\mathbb{P} = \mathbb{P}^r$ then the following properties are well known.

(i) for any $n \in \mathbb{Z}$ $\mathcal{O}_{\mathbb{P}}(n)$ is an invertible sheaf;

(ii) an invertible sheaf $\mathcal{O}_{\mathbb{P}}(n)$ is ample;

(iii) the homorphism of multiplication $S(n) \otimes S(m) \to S(n+m)$ induces the isomorphism $\mathcal{O}_{\mathbb{P}}(n) \otimes \mathcal{O}_{\mathbb{P}}(m) \simeq \mathcal{O}_{\mathbb{P}}(n+m)$;

(iv) for any graded S-module M and $n \in \mathbb{Z}$

$$\underline{M(n)} \simeq \underline{M} \otimes_{\mathcal{O}_{\mathbb{P}}} \mathcal{O}_{\mathbb{P}}(n)$$

None of these properties is valid for general $\mathbb{P}(Q)$.

1.5.1. Let $Q = \{1,1,2\}$. The restriction of $\mathcal{O}_{\mathbb{P}}(1)$ to $D_+(T_2)$ is given by the

$S_{(T_2)}$-module

$$S(1)_{(T_2)} = \{\frac{a}{T_2^k}:\ a \in S_{2k-1}\}\ .$$

It is clear that $S(1)_{(T_2)} = S_{(T_2)}\frac{T_0}{T_2} + S_{(T_2)}\frac{T_1}{T_2}$ is not a free $S_{(T_2)}$-module of rang one.

This is a counterexample to property (i).

1.5.2. On a weighted projective line $\mathbb{P}(q_0,q_1)$ all sheaves $\mathcal{O}_{\mathbb{P}}(n)$ are invertible. In fact, $\mathcal{O}_{\mathbb{P}}(n)|D_+(T_i)$ is associated to the $S_{(T_i)}$-module $S(n)_{(T_i)}$, freely generated by T_j^p/T_i^k, where $n = kq_i - pq_j$ and k/n, p/n are integers coprime with q_j and q_i respectively.

Since $\mathbb{P}(q_0,q_1) \simeq \mathbb{P}^1$ (1.3.1), an invertible sheaf $\mathcal{O}_{\mathbb{P}}(n)$ is equal to some $\mathcal{O}_{\mathbb{P}^1}(b_n)$. Moreover, if $\Gamma(\mathbb{P},\mathcal{O}_{\mathbb{P}}(n)) \neq 0$, then

$$b_n = \dim_k \Gamma(\mathbb{P},\mathcal{O}_{\mathbb{P}}(n)) - 1\ .$$

Thus, $\mathcal{O}_{\mathbb{P}}(n)$ is ample if $\dim_k \Gamma(\mathbb{P},\mathcal{O}_{\mathbb{P}}(n)) \geq 2$. But, if $n < \min\{q_0,q_1\}$ and $n > 0$, $\Gamma(\mathbb{P},\mathcal{O}_{\mathbb{P}}(n)) = 0$ (1.4.1).

This is a counterexample to property (ii).

1.5.3. In notations of 1.5.2 assume that $q_1 = q_0 + 1$, $q_0 > 1$. Then $b_{q_0} = b_{q_0+q_1+1} = 0$, $b_{q_1+1} < 0$. But

$$\mathcal{O}_{\mathbb{P}}(q_0) \otimes \mathcal{O}_{\mathbb{P}}(q_1+1) = \mathcal{O}_{\mathbb{P}^1}(b_{q_0}) \otimes \mathcal{O}_{\mathbb{P}^1}(b_{q_1+1}) \simeq \mathcal{O}_{\mathbb{P}^1}(b_{q_0}+b_{q_1+1})$$

$$\mathcal{O}_{\mathbb{P}}(q_0+q_1+1) \simeq \mathcal{O}_{\mathbb{P}^1}(b_{q_0+q_1+1})\ .$$

This is a counterexample to property (iii).

1.5.4. To obtain a counterexample to property (iv) we can take $M = S(m)$, note that $S(m)(n) = S(m+n)$ and use the counterexample from 1.5.3.

1.5.5. We refer to the paper of Delorme ([7]) for more details concerning properties of the sheaves $\mathcal{O}_{\mathbb{P}}(n)$. For example, one can find there a generalization of the duality theorem for $\mathbb{P}(Q)$, the particular case of which we have proved in 1.4.1.

We remark also that according to Mori ([19]) everything is well in the open set $V = \bigcap_{k>1} D_k$, where $D_k = \bigcup_{k \nmid q_i} D_+(T_i)$. Namely, V is the maximal open subscheme such that $\mathcal{O}_{\mathbb{P}}(1)|V$ is invertible and $(\mathcal{O}_{\mathbb{P}}(1)|V)^{\otimes m} \simeq \mathcal{O}_{\mathbb{P}}(m)|V$, $\forall m \in \mathbb{Z}$.

2. Bott's theorem.

2.1. Sheaves $\bar{\Omega}^i_{\mathbb{P}}$.

2.1.1. Let Ω_S^1 be the S-module of k-differentials of S. This is a free module with a basis $dT_0,\ldots,dT_r$. Denote by Ω_S^i its exterior i^{th} power $\overset{i}{\Lambda}(\Omega_S^1)$ (as usual we put $\Omega_S^0 = S$). This is a free S-module with the basis $\{dT_{s_1}\wedge\ldots\wedge dT_{s_i} : 0 \le s_1 < \ldots < s_i \le r\}$. Grade Ω_S^i by the condition

$$\deg(dT_{s_1}\wedge\ldots\wedge dT_{s_i}) = q_{s_1} + \ldots + q_{s_i} .$$

We have an isomorphism of graded S(Q)-modules

$$\Omega_S^i \simeq \bigoplus_{0\le s_1<\ldots<s_i\le r} S(-q_{s_1} - \cdots - q_{s_i}) .$$

For $i = r$ we obtain

$$\Omega_S^{r+1} \simeq S(-|Q|)$$

Let $d:S \to \Omega_S^1$ be the canonical universal differentiation. By definition of the partial derivatives we have

$$da = \sum_{j=0}^{r} \frac{\partial a}{\partial T_j}\, dT_j , \quad a \in S .$$

The k-linear map d extends to the exterior differentiation

$$d : \Omega_S^i \to \Omega_S^{i+1}$$

uniquely determined by the conditions

$$d(w\wedge w') = dw\wedge w' + (-1)^i w\wedge dw' , \quad w \in \Omega_S^i , \quad w' \in \Omega_S^j$$

$$d(d(w)) = 0 , \quad \forall w \in \Omega_S^i .$$

2.1.2. Recall the <u>Euler formula</u>:

$$na = \sum_{j=0}^{r} \frac{\partial a}{\partial T_j} q_j T_j , \quad \forall a \in S(Q)_n .$$

Using the linearity of both sides of this identity we may verify this formula only in the case when a is a monomial $T_0^{s_0}\ldots T_r^{s_r}$. But in this case it can be done without any difficulties.

2.1.3. Define the homomorphism of grades S-modules

$$\Delta : \Omega_S^i \to \Omega_S^{i-1} , \quad i \ge 1$$

by the formula

$$\Delta(dT_{s_1}\wedge\ldots\wedge dT_{s_i}) = \sum_{k=1}^{i} (-1)^{k+1} q_{s_k} T_{s_k} dT_{s_1}\wedge\ldots\wedge \widehat{dT_{s_k}} \wedge\ldots\wedge dT_{s_i}$$

<u>Lemma</u>.

(i) $\Delta(w\wedge w') = \Delta(w)\wedge w' + (-1)^i w\wedge\Delta(w')$, $w \in \Omega_S^i$, $w' \in \Omega_S^j$;

(ii) $\Delta(da) = na$, $a \in S_n$;

(iii) $\Delta(dw) + d(\Delta(w)) = nw$, $w \in (\Omega_S^i)_n$.

Using the linearity of Δ we may verify (i) only in the case $w = dT_{s_1}\wedge\ldots\wedge dT_{s_i}$, $w' = dT_{s'_1}\wedge\ldots\wedge dT_{s'_j}$. But this is easy.

Property (ii) is a corollary of the Euler formula. To verify property (iii) it suffices to consider the case $w = a\,dT_{s_1}\wedge\ldots\wedge dT_{s_i}$, $a \in S_k$. We have

$$\Delta(dw) = \Delta(da\wedge dT_{s_1}\wedge\ldots\wedge dT_{s_i}) = \Delta(da)\wedge dT_{s_1}\wedge\ldots\wedge dT_{s_i} - da\wedge\Delta(dT_{s_1}\wedge\ldots\wedge dT_{s_i}) =$$

$$= ka\,dT_{s_1}\wedge\ldots\wedge dT_{s_i} - da\wedge\Delta(dT_{s_1}\wedge\ldots\wedge dT_{s_i})$$

$$d(\Delta(w)) = d(a\Delta(dT_{s_1}\wedge\ldots\wedge dT_{s_i}) = da\wedge\Delta(dT_{s_1}\wedge\ldots\wedge dT_{s_i}) + a\,d(\Delta(dT_{s_1}\wedge\ldots\wedge dT_{s_i}) =$$

$$= da\wedge\Delta(dT_{s_1}\wedge\ldots\wedge dT_{s_i}) + a\,d(\sum_{\ell=1}^{i}(-1)^{\ell+1}q_{s_\ell}T_{s_\ell}dT_{s_1}\wedge\ldots\wedge\widehat{dT_{s_\ell}}\wedge\ldots\wedge dT_{s_i}) =$$

$$= da\wedge\Delta(dT_{s_1}\ldots dT_{s_i}) + (\sum_{\ell=1}^{i} q_{s_\ell})a\,dT_{s_1}\wedge\ldots\wedge dT_{s_i} .$$

Adding we get

$$\Delta(dw) + d(\Delta(w)) = (k + \sum_{\ell=1}^{i} q_{s_\ell})w = nw .$$

2.1.4. It is easy to identify the sequence

$$0 \to \Omega_S^{r+1} \xrightarrow{\Delta} \Omega_S^r \to \cdots\cdot \to \Omega_S^1 \to S$$

with the Koszul complex for the regular sequence $(q_0T_0,\ldots,q_rT_r)$. Thus, we obtain that it is an exact sequence.

Now put

$$\overline{\Omega}_S^i = \mathrm{Ker}(\Omega_S^i \xrightarrow{\Delta} \Omega_S^{i-1}) = \mathrm{Im}(\Omega_S^{i+1} \xrightarrow{\Delta} \Omega_S^i)$$

with the induced grading.

So, we have the exact sequences of graded S-modules:

$$0 \to \overline{\Omega}_S^i(n) \to \Omega_S^i(n) \to \overline{\Omega}_S^{i-1}(n) \to 0 , \quad i \geq 1 , \quad n \in \mathbb{Z}$$

2.1.5. Define the sheaf $\overline{\Omega}_{\mathbb{P}}^i$ on $\mathbb{P}(Q)$ by

$$\overline{\Omega}_{\mathbb{P}}^i = \underline{\overline{\Omega}_S^i}$$, where $\underline{M}$ denotes the sheaf associated to a graded S-module M.

Also we put

$$\overline{\Omega}_{\mathbb{P}}^i(n) = \underline{\overline{\Omega}_S^i(n)} , \quad n \in \mathbb{Z} .$$

Since $M \to \underline{M}$ is an exact functor we have exact sequences of sheaves on $\mathbb{P}(Q)$:

$$0 \to \overline{\Omega}_{\mathbb{P}}^i(n) \to \underline{\Omega_S^i(n)} \to \overline{\Omega}_{\mathbb{P}}^{i-1}(n) \to 0 , \quad i \geq 1 .$$

It is clear that for $i = r+1$, $\overline{\Omega}_{\mathbb{P}}^i(n) = 0$, thus for $i = r$

$$\overline{\Omega}_{\mathbb{P}}^r(n) = \underline{\Omega_S^{r+1}(n)} = \underline{S(n-|Q|)} = \mathcal{O}_{\mathbb{P}}(n-|Q|) .$$

2.1.6. Note that in the case $\mathrm{char}(k) = 0$ property (iii) in lemma 2.1.3 gives an algebraic proof of the acyclicity of the De Rham complex

$$0 \to k \to S \xrightarrow{d} \Omega_S \xrightarrow{d} \cdots \to \Omega_S^{r+1} \to 0 .$$

2.2. Justifications.

In this section we try to convince the reader that the sheaves $\overline{\Omega}^i_{\mathbb{P}}$ introduced in the previous section are good substitutes for the sheaves of germs of differentials $\Omega^i_{\mathbb{P}^r}$ on the usual projective space $\mathbb{P}^r$.

2.2.1. Let $\mathbb{P} = \mathbb{P}^r$. Let us show that

$$\overline{\Omega}^i_{\mathbb{P}}(n) = \Omega^i_{\mathbb{P}^r}(n) .$$

In this case

$$U = V(\mathcal{O}_{\mathbb{P}^r}(-1))^* = \mathrm{Spec}(\bigoplus_{n\in\mathbb{Z}} \mathcal{O}_{\mathbb{P}^r}(n))$$

is the complement to the zero section of the tautological line bundle $V(\mathcal{O}_{\mathbb{P}^r}(-1))$ on $\mathbb{P}^r$ and the canonical morphism $p : U \to \mathbb{P}^r$ is smooth.

The standard exact sequence

$$0 \to p^*\Omega^1_{\mathbb{P}} \to \Omega^1_U \to \Omega^1_{U/\mathbb{P}} \to 0$$

induces the exact sequences

$$0 \to p^*\Omega^i_{\mathbb{P}} \to \Omega^i_U \to \Omega^1_{U/\mathbb{P}} \otimes_{p^*} \Omega^{i-1}_{\mathbb{P}} \to 0 .$$

The homomorphism

$$\Delta : \Omega^1_S \to S \qquad (\sum_i a_i dT_i \to \sum_i a_i q_i T_i)$$

induces after restriction to U a surjective homomorphism of sheaves

$$\Delta : \Omega^1_U \to \mathcal{O}_U$$

(here we use that $(q_i, \mathrm{char}(k)) = 1$!). It is easy to verify that $\Delta(p^*\Omega^1_{\mathbb{P}}) = 0$ and hence Δ defines a surjective homomorphism

$$\tilde{\Delta} : \Omega^1_{U/\mathbb{P}} \to \mathcal{O}_U .$$

Since $\Omega^1_{U/\mathbb{P}}$ is invertible we obtain that $\tilde{\Delta}$ is in fact an isomorphism.

Thus we have exact sequences

$$0 \to p^*\Omega^i_{\mathbb{P}} \to \Omega^i_U \to p^*\Omega^{i-1}_{\mathbb{P}} \to 0$$

and applying p_* we obtain exact sequences

$$0 \to \bigoplus_{n\in\mathbb{Z}} \Omega^i_{\mathbb{P}}(n) \to \bigoplus_{n\in\mathbb{Z}} \underline{\Omega}^i_S(n) \to \bigoplus_{n\in\mathbb{Z}} \Omega^{i-1}_{\mathbb{P}}(n) \to 0 .$$

It is easy to see that in this way we obtain exact sequence 2.1.5 of the definition of $\overline{\Omega}^i_{\mathbb{P}}(n)$.

2.2.2. Lemma. In the notation of 1.3.2, let us assume that G is generated by pseudoreflections and its order is invertible in k. Then the canonical homomorphism

$$\Omega^i_{A/k} \to (\Omega^i_{B/k})^G$$

is an isomorphism of A-modules.

Proof. Since B (resp. A) is a polynomial algebra, the B-module $\Omega^i_{B/k}$ (resp. $\Omega^i_{A/k}$) is a free B-module (resp. A-module). Since B is a free A-module ([3],ch.5, 5, th.5), $\Omega^i_{B/k}$ is a free A-module. Let $\Omega^i_{A/k} \to \Omega^i_{B/k}$ be the canonical homomorphism of A-modules (the inverse image of a differential form). It is injective (because $\Omega^i_{A/k}$ is free and it is injective over a dense open subset of Spec A). Let T be its cokernel and

$$0 \to \Omega^i_{A/k} \to \Omega^i_{B/k} \to T \to 0$$

be the corresponding exact sequence.

Now, for every G-B-module M, the homomorphism $m \to \frac{1}{\#G} \sum_g g(m)$ is a projector onto a direct summand (here we use the assumption that $\#G$ is invertible in k), thus the functor $(\ )^G$ is exact. Applying this functor to the above exact sequence, we get an exact sequence

$$0 \to \Omega^i_{A/k} \to (\Omega^i_{B/k})^G \to T^G \to 0$$

where $(\Omega^i_{B/k})^G$, being a direct summand of a free A-module, is a projective A-module. This shows that dim. proj. $(T^G) \leq 1$ and, hence, depth $(T^G) \geq \dim B - 1$. This implies that $T^G = 0$ if its localization $(T^G)_P = 0$ for any prime P of A of height 1. Let Q be a prime ideal of B such that $Q \cap A = P$ and $G_Q = \{g \in G : g(Q) = Q\}$ be the decomposition group of Q. Then $(T^G)_P = (T_Q)^{G_Q} = \operatorname{Coker}(\Omega^i_{A_P/k} \to (\Omega^i_{B_Q/k})^{G_Q})$. Let G'_Q be the inertia group of Q, the subgroup of G_Q of elements which act trivially in the residue field K of B_Q. Then $B_Q \supset B'_Q = (B_Q)^{G'_Q} \supset (B_Q)^{G_Q} = A_P$, the extension $B'_Q \supset A_P$ is etale, the group G'_Q is a cyclic group of order e equal to the ramification index of the extension $B_Q \supset B'_Q$ ([3],ch.V, 5, n°5). This shows that $\Omega^i_{A_P/k} \otimes_{A_P} A'_Q \simeq \Omega^i_{B'_Q}$ and, hence, $\Omega^i_{A_P} = (\Omega^i_{B'_Q})^{G'_Q/G'_Q}$. Thus, it suffices to show that

$$\Omega^i_{B'_Q} = (\Omega^i_{B_Q})^{G'_Q} .$$

Passing to the completions, we may assume that $B_Q = K[[T]]$, $B'_Q = K[[T^e]]$ and a generator g of G'_Q acts on B_Q by multiplying T by a primitive e-th root of unity ζ. Let $t_1,\dots,t_{n-1}$ be a separable transcendence basis of K over k. Then

$$\Omega^i_{B'_Q} = \Sigma B'_Q \, dt_{j_1} \wedge \dots \wedge dt_{j_{i-1}} \quad dT^e$$

$$\Omega^i_{B_Q} = \Sigma B_Q \, dt_{j_1} \wedge \ldots \wedge dt_{j_{i-1}} \wedge dT$$

A direct computation shows that $(\Omega^i_{B_Q/k})^{G'_Q} = \Omega^i_{B'_Q/k}$.

2.2.3. Let $a : \mathbb{P}^r \to \mathbb{P}$ be the natural projection $\mathbb{P}^r \to \mathbb{P}^r/\mu_Q = \mathbb{P}(Q)$ from 1.2.2. Let us show that

$$\bar{\Omega}^i_{\mathbb{P}} \simeq a^G_*(\Omega^i_{\mathbb{P}}) ,$$

where $G = \mu_Q$ and a^G_* is the functor of invariant direct image ([11],5.1).

The action of G on $\mathbb{P}^r$ is induced by one on S. Since S^G is a polynomial algebra, this latter action is generated by pseudoreflections and hence, by lemma 2.2.2, we have an isomorphism of $S(Q)$-modules

$$\Omega^i_{S(Q)} \simeq (\Omega^i_S)^G$$

and, hence, an isomorphism of sheaves

$$\underline{\Omega}^i_{S(Q)} \simeq a^G_*(\underline{\Omega}^i_S) .$$

Applying a^G_* to the exact sequence (see 2.1.5)

$$0 \to \Omega^i_{\mathbb{P}^r} \to \underline{\Omega}^i_S \to \Omega^{i-1}_{\mathbb{P}^r} \to 0$$

and using the exactness of a^G_* (p is an affine morphism and $(\ )^G$ is an exact functor) we obtain an exact sequence:

$$0 \to a^G_*(\Omega^i_{\mathbb{P}^r}) \to \Omega^i_{S(Q)} \to a^G_*(\Omega^{i-1}_{\mathbb{P}^r}) \to 0 .$$

Since $a^G_*(\mathcal{O}_{\mathbb{P}^r}) = \mathcal{O}_{\mathbb{P}}$ we obtain by induction that $a^G_*(\Omega^i_{\mathbb{P}^r}) \simeq \bar{\Omega}^i_{\mathbb{P}}$.

2.2.4. Let us show that $\bar{\Omega}^i_{\mathbb{P}}$ coincides with the sheaf $\tilde{\Omega}^i_{\mathbb{P}}$ introduced for any V-variety in [23].

Recall that

$$\tilde{\Omega}^i_{\mathbb{P}} = j_*(\Omega^i_W)$$

where $j : W \to \mathbb{P}$ is the open immersion of the smooth locus of $\mathbb{P}$. In notations of 2.2.3, let us consider a commutative diagram

$$\begin{array}{ccc} a^{-1}(W) & \xrightarrow{\ j'\ } & \mathbb{P}^r \\ \downarrow{\scriptstyle a'} & & \downarrow{\scriptstyle a} \\ W & \xrightarrow{\ j\ } & \mathbb{P} \end{array}$$

Here $a' = a|a^{-1}(W)$ and j' is the natural immersion. Since W is smooth, the action of μ_Q on $a^{-1}(W)$ is generated locally by pseudoreflections. Then, by lemma 2.2.2, we get

$$\Omega^i_W = a'^G_*(\Omega^i_{a^{-1}(W)}) .$$

Since $\operatorname{codim}(\mathbb{P}-W,\mathbb{P}) \geq 2$ ($\mathbb{P}$ is a normal scheme) and $\mathbb{P}^r$ is smooth,

$$j'_*(\Omega^i_{a^{-1}(W)}) \simeq \Omega^i_{\mathbb{P}^r} .$$

Thus

$$\tilde{\Omega}^i_{\mathbb{P}} = j_*(\Omega^i_W) = j_*(a'^G_*(\Omega^i_{a^{-1}(W)}) = a^G_*(j'_*(\Omega^i_{a^{-1}(W)}) = a^G_*(\Omega^i_{\mathbb{P}^r})$$

2.3. Cohomology of $\overline{\Omega}^i_{\mathbb{P}}(n)$.

2.3.1. Let us consider the graded $S(Q)$-modules $\overline{\Omega}^i_S$, introduced in 2.1.4 and let $H^i_{\{m\}}$ denote the local cohomology group for a S-module M (cf. 1.4).

Proposition.

$$H^j_{\{m\}}(\overline{\Omega}^i_S) = \begin{cases} 0, & j \neq i+1,\ r+1 \\ k, & j = i+1 \neq r+1 . \end{cases}$$

Proof. We have exact sequences (2.1.4)

$$0 \to \overline{\Omega}^i_S \to \Omega^i_S \to \overline{\Omega}^{i-1}_S \to 0, \qquad i > 1$$

which, after applying the functor $H^i_{\{m\}}$, yield the exact sequences of local cohomology

$$\cdots \to H^{j-1}_{\{m\}}(\Omega^i_S) \to H^{j-1}_{\{m\}}(\overline{\Omega}^{i-1}_S) \to H^j_{\{m\}}(\overline{\Omega}^i_S) \to H^j_{\{m\}}(\Omega^i_S) \to \cdots .$$

Since $\Omega^i_S = S(-n)$ for some $n \in \mathbb{Z}$ and S is a Cohen-Macaulay ring, $H^j_{\{m\}}(\Omega^i_S) = 0$ if $j \neq r+1$. Thus, we have an isomorphism

$$H^j_{\{m\}}(\overline{\Omega}^i_S) \simeq H^{j-1}_{\{m\}}(\overline{\Omega}^{i-1}_S) \quad \text{for } j \neq r+1 .$$

By induction, we obtain

$$H^j_{\{m\}}(\overline{\Omega}^i_S) \simeq H^{j-i+1}_{\{m\}}(\overline{\Omega}_S) .$$

Now, first terms of the Koszul complex from (2.1.4) give an exact sequence

$$0 \to \overline{\Omega}^1_S \to \Omega^1_S \to m \to 0 ,$$

which easily implies that

$$H^1_{\{m\}}(\overline{\Omega}^1_S) = \begin{cases} 0, & 1 \neq 2,\ r+1 \\ k, & 1 = 2 \neq r+1 . \end{cases}$$

This proves the proposition.

Corollary. $\overline{\Omega}^i_S$ is a Cohen-Macaulay S-module if and only if $i = r$.

2.3.2. For any subset $J \subset [0,r] = \{0,\ldots,r\}$ denote by $|Q_J|$ the sum $\sum_{j\in J} q_j$. Notice that $|Q_{[0,r]}| = |Q|$ in our old notations. Put $a_n = \dim_k S(Q)_n$.

<u>Theorem</u>. Let $h(j,i;n) = \dim_k H^j(\mathbb{P},\bar{\Omega}^i_{\mathbb{P}}(n))$. Then

$$h(0,i;n) = \sum_{\#J=i} a_{n-|Q_J|} - h(0,i-1;n)\,, \quad i \geq 1\,, \quad n \in \mathbb{Z}$$

$$h(j,i;n) = 0\,, \quad \text{if } j \neq 0,i,r\,, \quad n \in \mathbb{Z}$$

$$h(i,i;0) = 1\,, \quad i = 0,\ldots,r$$

$$h(i,i;n) = 0\,, \quad n \neq 0\,, \quad i \neq r,0$$

$$h(r,i;n) = \sum_{\#J=r+1-i} a_{-n-|Q_J|} - h(r,i-1;n)\,, \quad i \geq 0\,, \quad n \in \mathbb{Z}$$

<u>Proof</u>. Using the same arguments as in the proof of theorem 1.4.1 we obtain the exact sequence

$$0 \to H^0_{\{m\}}(\bar{\Omega}^i_S) \to \bar{\Omega}^i_S \to \bigoplus_{n\in\mathbb{Z}} H^0(\mathbb{P},\bar{\Omega}^i_{\mathbb{P}}(n)) \to H_{\{m\}}(\bar{\Omega}^i_S) \to 0$$

and an isomorphism

$$H^j_{\{m\}}(\bar{\Omega}^i_S) \to \bigoplus_{n\in\mathbb{Z}} H^{j-1}(\mathbb{P},\bar{\Omega}^i_{\mathbb{P}}(n))\,.$$

Applying 2.3.1 we get that

$$H^0(\mathbb{P},\bar{\Omega}^i_{\mathbb{P}}(n)) = (\bar{\Omega}^i_S)_n = \operatorname{Ker}(\Omega^i_S \xrightarrow{\Delta} \bar{\Omega}^{i-1}_S)_n$$

$$H^j(\mathbb{P},\bar{\Omega}^i_{\mathbb{P}}(n)) = 0\,, \quad j \neq 0,i,r\,, \quad n \in \mathbb{Z}$$

$$H^i(\mathbb{P},\bar{\Omega}^i_{\mathbb{P}}(n)) = k\,, \quad n = 0\,, \quad i \neq r$$

$$H^i(\mathbb{P},\bar{\Omega}^i_{\mathbb{P}}(n)) = 0\,, \quad n \neq 0\,, \quad i \neq 0,r\,.$$

Now $\Omega^i_S = \bigoplus_{\#J=i} S(-|Q_J|)$ and Δ is surjective (2.1). So, we get all the assertions except the last one.

Consider exact sequence 2.1.5

$$0 \to \bar{\Omega}^i_{\mathbb{P}}(n) \to \underline{\Omega}^i_S(n) \to \Omega^{i-1}_{\mathbb{P}}(n) \to 0$$

and the corresponding cohomology sequence

$$H^{r-1}(\underline{\Omega}^i_S(n)) \to H^{r-1}(\bar{\Omega}^{i-1}_{\mathbb{P}}(n)) \to H^r(\bar{\Omega}^i_{\mathbb{P}}(n)) \to H^r(\underline{\Omega}^i_S(n)) \to H^r(\Omega^{i-1}_{\mathbb{P}}(n)) \to 0\,.$$

Since $\underline{\Omega}^i_S(n) \simeq \bigoplus_{\#J=i} \mathcal{O}_{\mathbb{P}}(n-|Q_J|)$ we can apply theorem 1.4.1 and obtain that

$$\dim_k H^r(\underline{\Omega}^i_S(n)) = \sum_{\#J=r+1-i} a_{-n-|Q_J|}\,.$$

Using this sequence and preceeding results we obtain the last equality.

2.3.3. <u>Corollary</u>.

$$H^0(\overline{\mathbb{P}},\overline{\Omega}^i_{\mathbb{P}}(n)) = 0\,, \qquad \text{if} \quad n < \min\{|Q_J| : \#J = i\}$$

$$H^r(\mathbb{P},\overline{\Omega}^i_{\mathbb{P}}(n)) = 0\,, \qquad \text{if} \quad n > -\min\{|Q_J| : \#J = r+1-i\}\,.$$

2.3.4. Corollary (Bott-Steenbrink). If $n > 0$ then

$$H^j(\mathbb{P},\overline{\Omega}^i_{\mathbb{P}}(n)) \neq 0$$

only when $j = 0$ and $n > \min\{|Q_J| : \#J = i\}$.

2.3.5. Corollary.

$$h(0,i;n) = \sum_{\ell=0}^{i} (-1)^{\ell+i} \sum_{\#J=\ell} a_{n-|Q_J|}$$

$$h(r,i;n) = h(0,r-i;-n)\,.$$

Here the first assertion immediately follows from 2.3.2 and to verify the second one we have to consider the identity

$$h(r,i;n) - h(0,r-i;-n) = h(0,r+1;-n) = \dim_k H^0(\mathbb{P},\overline{\Omega}^{r+1}_{\mathbb{P}}(-n)) = 0\,.$$

2.3.6. Corollary. If $k = \mathbb{C}$, then

$$H^i(\mathbb{P},\mathbb{C}) = \begin{cases} \mathbb{C}\,, & i \text{ even} \\ 0\,, & i \text{ odd} \end{cases}$$

$$h^{p,q}(\mathbb{P}) = \begin{cases} 1\,, & p = q \\ 0\,, & p \neq q\,. \end{cases}$$

This follows from the degeneracy of the spectral sequence $E_1^{p,q} = H^q(\mathbb{P},\overline{\Omega}^p_{\mathbb{P}}) \Rightarrow H^{p+q}(X,\mathbb{C})$ proven by Steenbrink [23].

3. Weighted complete intersections.

3.1. Quasicones.

3.1.1. Let X be a closed subscheme of a weighted projective space $\mathbb{P}(Q)$ and $p : U \to \mathbb{P}(Q)$ be the canonical projection.

The scheme closure of $p^{-1}(X)$ in A^{r+1} is called the affine quasicone over X. The point $0 \in C_X$ is called the vertex of C_X.

Let J be the Ideal of X in $\mathbb{P}$ then the ideal I of C_X in S is equal to $H^0(U,p^*(J)\otimes_{\mathcal{O}_{\mathbb{P}}} \mathcal{O}_U) = \bigoplus_{n\in\mathbb{Z}} H^0(\mathbb{P},J_{\mathcal{O}_{\mathbb{P}}} \otimes \mathcal{O}_{\mathbb{P}}(n))$.

3.1.2. Proposition.

(i) I is a homogeneous ideal of $S(Q)$;

(ii) the maximal ideal m_0 of the vertex of C_X coincides with the irrelevant ideal of the graded ring $S(Q)/I$ and has no immersed components (i.e. $\mathrm{depth}_m(S/I) \geq 1$);

(iii) The closed embedding $\mathrm{Proj}(S/I) \to \mathrm{Proj}(S) = \mathbb{P}$ corresponding to the natural projection $S \to S/I$ determines an isomorphism $\mathrm{Proj}(S/I) \simeq X$;

(iv) I is uniquely determined by the properties above.

This is an easy exercise in the theory of projective schemes, which we omit (it will not be used in the sequel).

3.1.3. An affine variety V is called quasiconic (or quasicone) if there is an effective action of G_m on V such that the intersection of the closures of all orbits is a closed point. This point is called the vertex of a quasicone.

3.1.4. Proposition. Let V be an affine algebraic variety over a field k. The following properties are equivalent:

(i) V is a quasicone;

(ii) $k[V] = \Gamma(V, O_V)$ has a nonnegative grading with $k[V]_0 \simeq k$;

(iii) there is a closed embedding $j : V \to \mathbb{A}^{r+1}$ such that $j(V)$ is invariant with respect to the action of G_m on $\mathbb{A}^{r+1}$ defined as in 1.2.1;

(iv) there is a closed embedding $j : V \to \mathbb{A}^{r+1}$ such that the ideal of $j(V)$ is generated by weighted-homogeneous polynomials with integer positive weights (i.e. homogeneous elements of some $S(Q)$).

The proof consists of standard arguments of the algebraic group theory (cf. [8], [20]).

Corollary. Any affine quasicone is a quasicone. Conversely any quasicone without immersed components in its vertex is an affine quasicone for some $X \subset \mathbb{P}(Q)$.

3.1.5. A closed subscheme $X \subset \mathbb{P}(Q)$ is called quasismooth (with respect to the embedding $X \to \mathbb{P}(Q)$) if its affine quasicone is smooth outside its vertex.

3.1.6. Theorem. A quasismooth closed subscheme $X \subset \mathbb{P}(Q)$ is a V-variety.

Proof. Let C_X be the affine quasicone over X and $x \in X$ be a closed point. In notations of the proof of 1.3.2 let $W_i = V_i \cap C_X$. Let us show that for any $y \in W_i$ over $x \notin W_i$ is nonsingular in y. We have to show that the tangent space $T_{C_X}(y)$ is not contained in the tangent space $T_{V_i}(y)$. Let $p' : C_X^* \to X$ be the restriction of p to $C_X^* = C_X - \{0\}$ and $F = p'^{-1}(x)_{red}$. The fibre F is an orbit of the point y with respect to the action of G_m restricted to C_X^*. If $(\overline{y}_0,\dots,\overline{y}_{i-1},1,\overline{y}_{i+1},\dots,\overline{y}_r)$ denotes the coordinates of y, then F coincides with the image of the map $G_m = \mathrm{Spec}(k[t,t^{-1}]) \to \mathbb{A}^{r+1} = \mathrm{Spec}(S)$ which is given by the formula:

$$(T_0,\dots,T_{i-1},T_i,T_{i+1},\dots,T_r) \to (\overline{y}_0 t^{q_0},\dots,\overline{y}_{i-1}t^{q_i-1},t^{q_i},\overline{y}_{i+1}t^{q_i+1},\dots,\overline{y}_r t^{q_2}).$$

The tangent line to the curve F is the image of the corresponding tangent map and defined by the equation

$$T_0 - \overline{y}_0 = q_0\overline{y}_0,\dots,T_{i-1} - \overline{y}_{i-1} = q_{i-1}\overline{y}_{i-1},T_i - 1 = q_i,\dots,T_r - \overline{y}_r = q_r\overline{y}_r .$$

It is clear that $T_F(y) \not\subset T_{V_i}(y) = V_i$ and, since $T_F(y) \subset T_{C_X}(y)$, we obtain that y is a nonsingular point of W_i.

The end of the proof is the same as in the proof of 1.3.2: we obtain that $U_k \subset X$ is locally isomorphic to the quotient of the nonsingular variety W_i by the isotropy group $G_y \subset G_m$ of the point y_i.

3.2. <u>Weighted complete intersections</u>.

3.2.1. Assume that the ideal $I \subset S$ of the affine quasicone C_X of $X \subset \mathbb{P}$ is generated by a regular sequence of homogeneous elements of the ring $S(Q)$. If $d_1,\dots,d_k$ are the degrees of these elements then we say that X is a <u>weighted complete intersection of multidegree</u> $\underline{d} = (d_1,\dots,d_k)$ and denote X by $V_{\underline{d}}(Q)$.

In case I is a principal ideal (F) and $F \in S(Q)_d$ we say that X is a <u>weighted hypersurface of degree</u> d and denote X by $V_d(Q)$.

3.2.2. In the sequel, C_X^* will denote the punctured affine quasicone $C_X - \{0\}$. Let $p : C_X^* \to X$ be the corresponding projection.

<u>Lemma</u>. Assume that $X = V_{\underline{d}}(Q)$ is quasismooth. Then

(i) $\mathrm{Pic}(C_X^*) = 0$ if $\dim X \geq 3$;

(ii) any G_m-equivariant etale covering of C_X^* is trivial if $\dim X \geq 2$;

(ii)' $\pi_1(C_X^*) = 0$ if $k = \mathbb{C}$ and $\dim X \geq 2$;

(iii) $H^i(C_X^*,\mathcal{O}_{C_X^*}) = 0$, $0 < i < \dim X$.

<u>Proof</u>. (i) Since the local ring $\mathcal{O}_{C_X,0}$ is a complete intersection ring of dimension 4, regular outside its maximal ideal, it is a factorial ring ([13], exp.XI). This shows that $\mathrm{Pic}(C_X^*) = \mathrm{Pic}(C_X)$. The latter group, being isomorphic to the group of classes of invertible divisorial ideals of a graded commutative ring, is trivial ([10]). This proves (i).

(ii) A similar reference ([13], exp.X) shows that $\mathcal{O}_{C_X 0}$ is pure. Hence, every etale covering of C_X^* is a restriction of an etale covering of C_X. Moreover, the same is true for G_m-equivariant coverings. Let $f: Y \to C_X$ be an irreducible G_m-equivariant etale covering of C_X. Then $Y = \mathrm{Spec}\, B$, where $B = \bigoplus_{n\in\mathbb{Z}} B_n$ is an integral $\mathbb{Z}$-graded k-algebra, and f is defined by an inclusion of graded rings $k[C_X] \subset B$. Let $m = k[C]_{>o} = \bigoplus_{n>o} k[C_X]_n$ be the maximal ideal of the vertex $o \in C_X$. Then $mB \subset B_{>o}$ and B/mB is a finite separable k-algebra. Since B is integral, this easily implies that $B_m = 0$ for $m < 0$ and B_o is a finite algebra over $k[C_X]/m = k$. Since B_o is a subalgebra of an integral algebra B, this implies that B_o is a field. Thus, we obtain that Y is a quasicone and its vertex is the only point lying over the vertex of C_X. Because f is etale and k is algebraically closed, this implies that f is an isomorphism. This proves (ii).

(ii)' Let C_X be G_m-equivariantly embedded into $\mathbb{C}^n$. The subgroup $\mathbb{R}_+$ of positive real numbers of the group $\mathbb{C}^*$ acts freely on C_X^*. Intersecting every $\mathbb{R}_+$-orbit with a sphere S_ε^{2n-1} of small radius ε with the center at the origin, we get a map $C_X^* \to \mathbb{R}_+ \times K_\varepsilon$, where $K_\varepsilon = S^{2n-1} \cap C_X$. It is easily verified that this map is a diffeomorphism of C_X^* onto $\mathbb{R}_+ \times K_\varepsilon$. Now, since the vertex of C_X is a complete intersection isolated singularity, the space K_ε is (d-2)-connected ($d = \dim C_X = \dim X + 1$) (see [14,18]). Thus, $\pi_1(C_X^*) \simeq \pi_1(K_\varepsilon) = 0$ if $\dim X \geq 2$.

To verify (iii) we again use the local cohomology theory. Since $C_X = \mathrm{Spec}(S/I)$ is affine,

$$H^i(C_X^*, \mathcal{O}_{C_X^*}) = H^{i+1}_{\{0\}}(C_X, \mathcal{O}_{C_X}) = H^{i+1}_{\{m_0\}}(S/I).$$

Since S/I, being a quotient of a regular ring by a regular sequence, is a Cohen-Macaulay ring, $H^{i+1}_{\{m_0\}}(S/I) = 0$, if $i+1 \neq \dim(S/I) = \dim X + 1$.

3.2.3. Remark. If $\mathrm{char}\, k > 0$, then $\pi_1^{alg}(C_X^*)$ may be not trivial. For example, $\pi_1^{alg}(\mathbb{A}^n - \{0\}) \neq 0$, because A^n has nontrivial etale coverings.

3.2.4. Theorem. Under the conditions of the lemma

(i) $\mathrm{Pic}(X) \simeq \mathbb{Z}$, if $\dim X \geq 3$;

(ii) $\pi_1^{alg}(X) = 0$, if $\dim X \geq 2$;

(ii)' $\pi_1(X) = 0$, if $k = \mathbb{C}$ and $\dim X \geq 2$;

(iii) $H^i(X, \mathcal{O}_X(n)) = 0$, $n \in \mathbb{Z}$, $0 < i < \dim X$.

Proof. Let L be an invertible sheaf on X. Since $\mathrm{Pic}(C_X^*) = 0$, $p^*(L) = \mathcal{O}_{C_X^*}$ and is determined as a G_m-sheaf by some character $\chi_L \in H^1(G_m, \underline{\mathrm{Aut}}(\mathcal{O}_{C_X^*})) = H^1(G_m, G_m) \simeq \mathbb{Z}$. In this way we obtain a homomorphism $f: \mathrm{Pic}(X) \to \mathbb{Z}$. If $p^*(L) = p^*(L')$ as G_m-sheaves, then $L = p_*^{G_m}(p^*(L)) = L' = p_*^{G}(p^*L'))$ and, hence, f is injective. This proves (i).

Let X' be an etale finite covering of X and A be a corresponding $\mathcal{O}_X$-Algebra (i.e. $X' = \mathrm{Spec}(A)$). Since the covering $\overline{X}' = X' \times_X C_X^*$ of C_X^* is a

G_m-equivariant covering, by lemma 3.2.2 (ii) we get that $\overline{X}' = \mathrm{Spec}(p^*(A))$ is trivial.

Hence the G_m-$\mathcal{O}_{C_X^*}$-Algebra $\mathcal{B} = p^*(A) = A_{\mathcal{O}_X} \otimes \mathcal{O}_{C_X^*}$ splits, i.e. $\mathcal{B} = \mathcal{B}_1 \times \mathcal{B}_2$, where $\mathcal{B}_i$ are nontrivial. Since G_m is connected, the Subalgebras $\mathcal{B}_1$ and $\mathcal{B}_2$ are invariant Subalgebras and we have a splitting of G_m-Algebras $\mathcal{B} = \mathcal{B}_1 \times \mathcal{B}_2$. Applying $p_*^{G_m}$, we obtain that $A = p_*^{G_m}(\mathcal{B}) = p_*^{G_m}(\mathcal{B}_1) \times p_*^{G_m}(\mathcal{B}_2)$ splits. This shows that the covering X' splits and proves (ii).

To prove (ii)' we apply Lemma 3.2.2 (ii) and notice that the canonical homomorphism $\pi_1(C_X^*) \to \pi_1(X)$ is surjective because the fibres of $C_X^* \to X$ are pathwise connected.

To prove (iii) we note that

$$H^i(C_X^*, \mathcal{O}_{C_X^*}) = H^i(C_X^*, \mathcal{O}_X \otimes_{\mathcal{O}_{\mathbb{P}}} \mathcal{O}_U) = H^i(X, \mathcal{O}_X \otimes p_* \mathcal{O}_U) = \bigoplus_{n \in \mathbb{Z}} H^i(X, \mathcal{O}_X \otimes_{\mathcal{O}_{\mathbb{P}}} \mathcal{O}_{\mathbb{P}}(n)) .$$

But $\mathcal{O}_X \otimes_{\mathcal{O}_{\mathbb{P}}} \mathcal{O}_{\mathbb{P}}(n) \simeq \mathcal{O}_X(n)$ and we can apply 3.2.2 and obtain (iii).

3.2.5. Remark. The proof of (i) easily gives that $\mathrm{Pic}(X)$ is generated by some $\mathcal{O}_X(n)$, where, in general, $n \neq 1$. For example, $\mathrm{Pic}(\mathbb{P}(1,\ldots,1,n))$ is generated by $\mathcal{O}_{\mathbb{P}}(2)$.

3.2.6. One can also prove 3.2.4 (and its generalizations to torical spaces) using the methods of [13] (cf. [9]).

3.3. The dualizing sheaf.

3.3.1. Recall that according to Grothendieck for any normal integral projective Cohen-Macaulay variety X there is a sheaf ω_X (the dualizing sheaf) such that

$$H^i(X,F) = (\mathrm{Ext}^{n-i}(X;F,\omega_X))^* \qquad (n = \dim X)$$

for any coherent $\mathcal{O}_X$-Module F. The sheaf ω_X can be determined as the sheaf of germs of differential forms which are regular at nonsingular points of X (see, for example [16]).

In other words,

$$\omega_X = j_*(\Omega_Z^n)$$

where $j : Z \to X$ is the open immersion of the nonsingular locus of X.

In this section we shall compute ω_X for a quasismooth weighted complete intersection.

3.3.2. Lemma. Let X be a closed quasismooth subscheme of $\mathbb{P}$, C_X be its projecting quasicone, Z be the nonsingular locus of X, then

$$p_*^{G_m}(\Omega^1_{C_X^*/X})|Z \simeq \mathcal{O}_Z .$$

Proof. The embedding of smooth schemes $C_X^* \to U$ defines an exact sequence

$$0 \to J/J^2 \to \Omega^1_{U/\mathbb{P}} \otimes_{\mathcal{O}_U} \mathcal{O}_{C_X^*} \to \Omega^1_{C_X^*/X} \to 0$$

where J is the ideal sheaf of C_X^* in U. Consider the surjective homomorphism $\tilde{\Delta} : \Omega^1_{U/\mathbb{P}} \to \mathcal{O}_U$ from 2.2.1 (its construction does not use the assumption that $\mathbb{P} = \mathbb{P}^r$, the latter is used only to check that $\tilde{\Delta}$ is an isomorphism). It is easy to see that the induced map $\overline{\Delta} : \Omega^1_{C_X^*/X} \to \mathcal{O}_{C_X^*}$ is well defined and is surjective.

Since $p_*^{G_m}$ is exact, the map

$$p_*^{G_m}(\overline{\Delta}) : p_*^{G_m}(\Omega^1_{C_X^*/X}) \to \mathcal{O}_X$$

is surjective. Thus, it is sufficient to show that the restriction of the left hand side sheaf to Z is an invertible sheaf.

This verification is local. Let $x \in Z$ and Z_x be its neighbourhood of the form W/G where W is a nonsingular subvariety of C_X^* of codimension 1 and G is a finite subgroup of G_m constructed in the proof of theorem 3.1.6.

Since W is regularly embedded in C_X^*, we have an exact sequence:

$$0 \to N_{W/C_X} \to \Omega^1_{C_X^*/X} \otimes_{\mathcal{O}_{C_X^*}} \mathcal{O}_W \to \Omega^1_{W/Z_x} \to 0 .$$

Since N_{W/C_X} is locally free of rank 1 we may assume (replacing W by smaller one) that $N_{W/C_X} \simeq \mathcal{O}_W$.

It is clear that

$$p_*^{G_m}(\Omega^1_{C_X^*/X} \otimes_{\mathcal{O}_{C_X}} \mathcal{O}_W) = p_*^G(\Omega^1_{C_X^*/X} \otimes_{\mathcal{O}_{C_X^*}} \mathcal{O}_W) .$$

Since x is nonsingular, G acts by pseudoreflections and hence $p_*^G(\Omega^1_{W/Z_X}) = 0$ (see the proof of 2.2.2). Applying $p_*^{G_m}$ to the above sequence we obtain

$$p_*^{G_m}(\Omega^1_{C_X^*/X})|Z_x = p_*^G(\mathcal{O}_W) = \mathcal{O}_{Z_x} .$$

This proves the lemma.

3.3.3. Proposition. In conditions of 3.3.2

$$\omega_X = p_*^{G_m}(\Omega^{n+1}_{C_X^*}) \qquad (n = \dim X) .$$

Proof. Since X is a normal Cohen-Macaulay variety (it follows easily from 3.1.6), by 3.3.1 it is sufficient to show that

$$p_*^{G_m}(\Omega^{n+1}_{C_X^*})|Z \simeq \Omega^n_Z .$$

Consider the exact sequence

$$0 \to p^*\Omega^1_Z \to \Omega^1_{C_X^*}|p^{-1}(Z) \to \Omega^1_{C_X^*}|p^{-1}(Z) \to 0 .$$

Applying $p_*^{G_m}$ and using 3.3.2 we obtain the exact sequence

$$0 \to \Omega_Z^1 \to p_*^{G_m}(\Omega_{C_X^*}^1)|Z \to \mathcal{O}_Z \to 0 .$$

Taking the exterior power we get

$$\Omega_Z^n \simeq \Omega_Z^n \otimes_{\mathcal{O}_Z} \mathcal{O}_Z \simeq p_*^{G_m}(\Omega_{C_X^*}^{n+1})|Z$$

q.e.d.

3.3.4. Theorem. Let $X = V_{\underline{d}}(Q)$ be a quasismooth weighted complete intersection of multidegree $\underline{d} = (d_1,\dots,d_s)$. Then

$$\omega_X \simeq \mathcal{O}_X(|\underline{d}| - |Q|)$$

where $|\underline{d}| = d_1+\cdots+d_s$.

Proof. Let I be the ideal of the projecting quasicone over X and $B = S/I$. There is an isomorphism of graded A-modules

$$I/I^2 = B(-d_1) + \cdots + B(-d_s) .$$

The exact sequence

$$0 \to I/I^2 \to \Omega_S^1 \otimes_S B \to \Omega_B^1 \to 0$$

gives the homomorphism

$$f : \Omega_B^{r+1-s}(-|\underline{d}|) = \overset{s}{\Lambda}(I/I^2) \otimes_B \overset{r+1-s}{\Lambda}(\Omega_B^1) \to \overset{2+1}{\Lambda}(\Omega_S^1) \otimes_S B = B(-|Q|) .$$

Since C_X^* is smooth, the restriction of f to C_X^* is an isomorphism. Hence

$$\Omega_{C_X^*}^{r+1-s} = B(|\underline{d}| - |Q|) .$$

It remains to use the above proposition.

3.4. The Poincare series.

3.4.1. Let $A = \bigoplus_{n\geq 0} A_n$ be a graded k-algebra of finite type. Then its Poincare series is defined by

$$P_A(t) = \sum_{n=0}^{\infty} (\dim_k A_n) t^n .$$

If $x_0,\dots,x_r$ are homogeneous generators of A and $q_0,\dots,q_r$ are its degrees, then $P_A(t)$ is a rational function of the form

$$P_A(t) = F(t) / \prod_{i=0}^{r} (1 - t^{q_i})$$

where $F(t)$ is a polynomial ([2], 11.1).

3.4.2. Assume that $A = S(Q)$ is a graded polynomial k-algebra. Then ([3], ch.V, §5, n°1)

$$P_{S(Q)}(t) = 1/\prod_{i=0}^{r}(1-t^{q_i}) .$$

Let $f_1,\dots,f_s$ be a regular sequence of homogeneous elements of the ring $S(Q)$ and $d_1,\dots,d_r$ be its degrees, let $A = S(Q)/(f_1,\dots,f_s)$. Then

$$P_A(t) = \prod_{i=1}^{s}(1-t^{d_i})/\prod_{i=0}^{r}(1-t^{q_i}) .$$

This formula follows from the above formula. Put $A^0 = S(Q)$, $A^i = S(Q)/(f_1,\dots,f_i)$. Then $A^i = A^{i-1}/(f_i)$ and obviously

$$t^{d_i}P_{A^{i-1}}(t) + P_{A^i}(t) = P_{A^{i-1}}(t) .$$

Thus

$$P_{A^i}(t) = (1-t^{d_i})P_{A^{i-1}}(t) , \qquad i = 1,\dots,s$$

and we obtain our formula.

3.4.3. For $X = \mathrm{Proj}(A)$ we put

$$P_X(t) = \sum_{n=0}^{\infty}(\dim_k H^0(X,\mathcal{O}_X(n)))t^n .$$

Lemma. Let $m = \bigoplus_{n>0} A_n$ be the irrelevant ideal of A. Assume that $\mathrm{depth}_m(A) \geq 2$ (for example, A is normal). Then

$$P_A(t) = P_X(t) .$$

The same argument as in the proof of 3.2.4 (iii) and 1.4.2(i) shows that the Serre homomorphism of graded algebras

$$A \to \bigoplus_{n\in\mathbb{Z}} H^0(X,\mathcal{O}_X(n))$$

is bijective.

3.4.4. Theorem. Let $X = V_{\underline{d}}(Q)$ be a quasismooth weighted complete intersection, $P_X(t) = \sum_{n=0}^{\infty} a_n t^n$ be the power series defined above. Then

$$P_X(t) = \prod_{i=1}^{s}(1-t^{d_i})/\prod_{i=0}^{r}(1-t^{q_i}) .$$

Corollary. Define $p_g(X) = \dim_k H^{\dim X}(X,\mathcal{O}_X)$, then in notations of the theorem

$$p_g(V_{\underline{d}}(Q)) = a_{|\underline{d}|-|Q|} .$$

Indeed, since $\omega_X = \mathcal{O}_X(|\underline{d}|-|Q|)$ is the dualizing sheaf (3.3.4) we have that $\dim_k H^{\dim X}(X,\mathcal{O}_X) = \dim_k \mathrm{Hom}(\mathcal{O}_X,\omega_X) = \dim_k H^0(X,\omega_X) = a_{|\underline{d}|-|Q|}$.

3.5. Examples.

3.5.1. We shall say that two closed subvarieties $X \subset \mathbb{P}$ and $X' \subset \mathbb{P}'$ are affine isomorphic if their affine quasicones are isomorphic and projectively isomorphic if their quasicones are G_m-isomorphic. It is clear that in general there are only two implications

projectively isomorphic $\Rightarrow$ affine isomorphic

projectively isomorphic $\Rightarrow$ isomorphic

between these three notions.

3.5.2. Weighted plane curves. A quasismooth hypersurface $X = V_d(Q)$ in a weighted projective plane $\mathbb{P}(q_0,q_1,q_2)$ is a smooth projective curve. Its dualizing sheaf coincides with the canonical sheaf Ω^1_X and we have (3.3.4):

$$\Omega^1_X = \mathcal{O}_X(d-q_0-q_1-q_2) .$$

Its genus is calculated by the formula

$$g = \text{coefficient at } t^{d-|Q|} \text{ in the formal series}$$

$$(1-t^d)/\prod_{i=0}^{2}(1-t^{q_i}) .$$

The affine quasicone of such a curve is given by a weighted-homogeneous equation $f(x_0,x_1,x_2) = 0$ with an isolated singularity at the origin. Such singularities were studied by many authors ([1,8,18,20]).

Let

$$m = d - q_0 - q_1 - q_2 .$$

Each weighted plane curve with $m < 0$ is affine isomorphic to one of the following curves

$\mathbb{P}(q_0,q_1,q_2)$	d	Equation	Name
$\mathbb{P}(1,1,1)$	1	$x_0 = 0$	
$\mathbb{P}(1,k,k)$	2k	$x_0^{2k} + x_1^2 + x_2^2 = 0$	A_{2k-1}, $k \geq 1$
$\mathbb{P}(2,2k+1,2k+1)$	4k+2	$x_0^{2k+1} + x_1^2 + x_2^2 = 0$	A_{2k}, $k \geq 1$
$\mathbb{P}(2,k-2,k-1)$	2k-2	$x_0^{k-1} + x_1^2 x_0 + x_2^2 = 0$	D_k, $k \geq 4$
$\mathbb{P}(3,4,6)$	12	$x_0^4 + x_1^3 + x_2^2 = 0$	E_6
$\mathbb{P}(4,6,9)$	18	$x_0^3 x_1 + x_1^3 + x_2^2 = 0$	E_7
$\mathbb{P}(6,10,15)$	30	$x_0^5 + x_1^3 + x_2^2 = 0$	E_8

The equations of corresponding projecting quasicones are well known two-dimensional singularities, which are called the platonic singularities, Du Val singularities, Klein singularities, ADE singularities, double rational singularities, simple singularities, O-modal singularities.

Note that any curve which is affine isomorphic to a curve of type D_k or E is projectively isomorphic to this curve.

It is clear that all such curves with $m < 0$ are isomorphic to $\mathbb{P}^1$.

When $m = 0$ each weighted plane curve is projectively isomorphic to one of the following curves:

$\mathbb{P}(q_0,q_1,q_2)$	d	Equation	Name
$\mathbb{P}(1,1,1)$	3	$x_0^3 + x_1^3 + x_2^3 + ax_0x_1x_2 = 0, \quad a^3 + 27 \neq 0$	$\tilde{E}_6$ or P_8
$\mathbb{P}(1,1,2)$	4	$x_0^4 + x_1^4 + x_2^2 + ax_0^2x_1^2 = 0, \quad a^2 - 4 \neq 0$	$\tilde{E}_7$ or X_9
$\mathbb{P}(1,2,3)$	6	$x_0^6 + x_1^3 + x_2^2 + ax_1^2x_2^2 = 0, \quad 4a^3 + 27 \neq 0$	$\tilde{E}_8$ or J_{10}

It can be shown (V. I. Arnold) that for any fixed m there is only a finite number of collections $(q_0,q_1,q_2;d)$ for which there is a smooth weighted plane curve $V_d(q_0,q_1,q_2)$.

For $m = 1$ there are exactly 31 collections. The corresponding affine quasicones have a canonical quasihomogeneous singularity embeddable in $\mathbb{A}^3$. There is a natural correspondence between the 31 collections and the 31 possible signatures of the Fuchsian groups of the first kind with compact quotient for which the algebra of automorphic forms is generated by three elements ([8,25]).

Of course, a general smooth projective curve is not isomorphic to any weighted plane curve.

3.5.3. Surfaces. There are no classification results in this case, there are only some interesting examples.

Let $f(x_0,x_1,x_2) = 0$ be an equation of a smooth weighted plane curve $V_d(Q)$. Then the equation

$$f(x_0,x_1,x_2) + x_3^d = 0$$

defines a quasismooth hypersurface $V_d(q_0,q_1,q_2,1)$.

For curves with $m = 0$ we obtain in this way dell Pezzo surfaces [15] of degree 3, 2 and 1 respectively (M. Reid).

For curves with $m = 1$ we obtain simply-connected projective surfaces with the dualizing sheaf $\omega_X \cong \mathcal{O}_X$. Resolving its singularities (which are double rational points) we get minimal models of nonsingular K3-surfaces. One example of such a surface is the following Klein surface:

$$V_{42}(6,14,21,1) : x_0^7 + x_1^3 + x_2^2 + x_3^{42} = 0 .$$

This surface has 3 singular points

$(1,-1,0,0)$ of type A_1
$(0,-1,1,0)$ of type A_6
$(-1,0,1,0)$ of type A_2 .

For any such surface the complement to the curve $x_3 = 0$ is isomorphic to the affine surface with an equation

$$f(x_0,x_1,x_2) + 1 = 0$$

which is diffeomorphic to the Milnor space F_θ for the singularity $f(x_0,x_1,x_2) = 0$ ([18]). This fact can be used for the explanation of some observations in the singularity theory by means of the theory of algebraic surfaces (see [21]).

3.5.4. Multiple spaces. Let $X \to \mathbb{P}^{r-1}$ be a finite Galois covering with a cyclic automorphism group of order m branched along a smooth surface $W \subset \mathbb{P}^{r-1}$ of degree d. Let $f(x_0,\dots,x_{r-1}) = 0$ be the equation of W. Assume that $(d,\mathrm{char}(k)) = 1$. Then X is isomorphic to a weighted quasismooth hypersurface

$$V_d(Q) : f(x_0,\dots,x_{r-1}) + x_r^m = 0$$

where $Q = \{1,\dots,1,d/m\}$.

It is easy to see that such X is smooth. From 3.2.4 we obtain that all such varieties are simply-connected if $r \geq 3$ (i.e. $\pi_1^{alg}(X) = 0$ or $\pi_1(X) = 0$ if $k = \mathbb{C}$) (cf. [22]). Moreover, $\mathrm{Pic}(X) \simeq \mathbb{Z}$ if $r \geq 4$.

The Poincare series $P_X(t)$ has the form (3.4.4):

$$P_X(t) = (1-t^d)/(1-t)^r(1-t^{d/m}) = (1+t^{d/m}+\dots+t^{d(m-1)/m})/(1-t)^r .$$

In particular,

$$p_g(X) = \text{the coefficient at } t^{d-r-d/m} = \sum_{s=0}^{m-1} \begin{pmatrix} \frac{d(m-1-s)}{m} - 1 \\ r-1 \end{pmatrix} .$$

For example

$m = 2$, $r = 2$ (hyperelliptic curve) $p_g = d/2 - 1$

$m = 2$, $r = 3$, $d = 6$ (K3-surface) $p_g = 1$.

It is very useful for the construction problems in algebraic geometry to consider also weighted multiple planes, cyclic coverings of weighted projective spaces. For example, the Klein surface from 3.5.3 is such a multiple plane.

4. The Hodge structure on the cohomology of weighted hypersurfaces.

4.1. A resolution of $\tilde{\Omega}_X^i$.

Let $X = V_N(Q)$ be a quasismooth weighted hypersurface, C_X its affine quasicone, $I \subset S(Q)$ the ideal of C_X, $f \in S(Q)_N$ its generator, $A = S(Q)/I$ the coordinate ring of C_X, m_0 the maximal ideal of the vertex of C_X, $C_X^* = C_X - \{0\}$.

Since X is a V-variety (3.1.6) its cohomology has (in case $k = \mathbb{C}$) a pure Hodge structure and the corresponding Hodge numbers are calculated by the formula (see [23])

$$h^{p,q}(X) = \dim_k(H^q(X,\tilde{\Omega}_X^p)) .$$

In this section we shall construct a suitable resolution for the sheaf $\tilde{\Omega}^p_X$.

4.1.1. Define a k-linear map

$$d_f : \Omega^i_S \to \Omega^{i+1}_S , \qquad i \geq 0$$

setting for homogeneous elements of the S-module Ω^i_S (2.1.1)

$$d_f(w) = fdw + (-1)^{i+1} \frac{|w|}{N} w \wedge df ,$$

where $|w|$ denotes for brevity the degree of w.

Lemma.

(i) $d_f(w \wedge w') = d_f(w) \wedge w' + (-1)^i w \wedge d_f(w')$, $w \in \Omega^i_S$, $w' \in \Omega^j_S$;

(ii) $d_f(d_f(w)) = 0$, $\forall w \in \Omega^i_S$;

(iii) $d(d_f(w)) = (1 + \frac{|w|}{N}) df \wedge dw$, $\forall w \in \Omega^i_S$;

(iv) $d_f(dw) = \frac{|w|}{N} df \wedge dw$, $\forall w \in \Omega^i_S$;

(v) $d_f((\Omega^i_S)_n) \subset (\Omega^{i+1}_S)_{n+N}$.

This is directly verified.

Let us show that d_f induces a linear map of S-modules $\overline{\Omega}^i_S = \mathrm{Ker}(\Omega^i_S \xrightarrow{\Delta} \Omega^{i-1}_S)$ (2.1.4).

Lemma (continuation).

(vi) $d_f(\Delta(w)) = -\Delta(d_f(w))$, $w \in \Omega^i_S$;

(vii) $d_f(\overline{\Omega}^i_S) \subset \overline{\Omega}^{i+1}_S$.

It is clear that (vii) follows from (vi). Let us prove (vi). Recalling properties of the map Δ (2.1.3), we obtain

$$\Delta(d_f(w)) = \Delta(fdw + (-1)^{i+1} \frac{|w|}{N} w \wedge df) = f\Delta(dw) + (-1)^{i+1} \frac{|w|}{N} \Delta(w \wedge df) =$$

$$= -fd(\Delta(w)) + f|w|w + (-1)^{i+1} \frac{|w|}{N} \Delta(w) \wedge df - |w| fw =$$

$$= -fd(\Delta(w)) + (-1)^i \frac{|\Delta(w)|}{N} \Delta(w) \wedge df = -d_f(\Delta(w)) .$$

4.1.2. Properties (ii) and (v) of the lemma make possible to introduce the following complex $R^{\bullet}_i$ of graded S-modules:

$$R^k_i = \Omega^k_S((k-i)N)$$

$$d_k = (-1)^k d_f : R^k_i \to R^{k+1}_i .$$

Property (vi) implies that the homomorphisms $\Delta : \Omega^k_S \to \Omega^{k-1}_S$ determine morphisms of complexes

$$\Delta : R^{\bullet}_i \to R^{\bullet}_{i-1}[-1] .$$

Property (vii) shows that

$$\bar{R}_i^\bullet = (\bar{\Omega}_S^k((k-i)N))_k$$

is a subcomplex of $R_i^\bullet$ such that

$$\bar{R}_i^\bullet = \mathrm{Ker}(R_i^\bullet \to R_{i-1}^\bullet[-1])$$
$$\bar{R}_{i-1}^\bullet[-1] = \mathrm{Im}(R_i^\bullet \to R_{i-1}^\bullet[-1]) .$$

Thus we have the exact sequence of complexes of graded S-modules:

$$0 \to \bar{R}_i^\bullet \to R_i^\bullet \to \bar{R}_{i-1}^\bullet[-1] \to 0 , \quad i \in \mathbb{Z} .$$

4.1.3. The multiplication by f defines the inclusion of graded S-modules

$$\Omega_S^k \to \Omega_S^k(N) , \quad \bar{\Omega}_S^k \to \bar{\Omega}_S^k(N) ,$$

which induces the inclusion of complexes

$$R_i^\bullet \to R_{i-1}^\bullet , \quad \bar{R}_i^\bullet \to \bar{R}_{i-1}^\bullet .$$

Consider the corresponding quotient complexes

$$K_i^\bullet = R_{i-1}^\bullet / R_i^\bullet , \quad \bar{K}_i^\bullet = \bar{R}_{i-1}^\bullet / \bar{R}_i^\bullet .$$

The exact sequence of complexes from 4.1.2 induces the exact sequence of complexes of graded S-modules:

$$0 \to \bar{K}_i^\bullet \to K_i^\bullet \to K_{i-1}^\bullet[-1] \to 0 .$$

4.1.4. Lemma (De Rham). Let A be a commutative ring, $w \in A^{r+1}$ be a regular sequence of elements of A, $h \in \overset{p}{\Lambda}(A^{r+1})$, $p \le r$. Then $w \wedge h = 0$ iff $\exists\, \beta \in \overset{p-1}{\Lambda}(A^{r+1})$ such that $h = w \wedge \beta$.

This is a reformulation of the theorem of acyclicity of the Koszul complex for a regular sequence.

We shall use this lemma in the following situation: A is the coordinate ring of C_X, $A^{r+1} = \Omega_S^1 / f\Omega_S^1$, w is the image of df in $\Omega_S^1 / f\Omega_S^1$.

Since C_X^* is smooth, the jacobian ideal

$$\theta_f = \left(\frac{\partial f}{\partial T_0}, \ldots, \frac{\partial f}{\partial T_0} \right) \subset S(Q)$$

is m_0-primary and hence df determines a regular sequence.

It is clear that the differential of the complex $K_i^\bullet$ coincides (up to the multiplication by a constant) with the exterior multiplication by df. Since

$$K_i^s = \overset{s}{\Lambda}(\Omega_S^1 / f\Omega_S^1)$$

we may use the De Rham lemma and deduce

Corollary:

$$H^q(K_i^\bullet) = 0 , \quad q \neq r+1 , \quad \forall i \in \mathbb{Z} .$$

4.1.5. Proposition.

$$H^q(\overline{K}_i^\bullet) = 0 , \quad q \geq 0 , \quad i \in \mathbb{Z} .$$

Proof. The above corollary and exact sequence 4.1.3 imply that

$$H^q(\overline{K}_i^\bullet) = H^{q-1}(\overline{K}_{i-1}^\bullet[-1]) = H^{q-2}(\overline{K}_{i-1}^\bullet) , \quad q \leq r .$$

Since for $q < 0$ and $q > r$ $H^q(\overline{K}_i^\bullet) = 0$ we obtain the assertion of the proposition.

4.1.6. Define a graded A-module $\overline{\Omega}_A^i$ by the equality

$$\overline{\Omega}_A^i = \mathrm{Ker}(\overline{K}_i^i \to \overline{K}_i^{i+1}) = \mathrm{Ker}(\overline{\Omega}_S^i(N)/\overline{\Omega}_S^i \xrightarrow{d_f} \overline{\Omega}_S^{i+1}(2N)/\overline{\Omega}_S^{i+1}(N)) .$$

Then we deduce from 4.1.5 that the sequence of graded A-modules

$$0 \to \overline{\Omega}_A^i \to \overline{\Omega}_S^i(N)/\overline{\Omega}_S^i \to \overline{\Omega}_S^{i+1}(2N)/\overline{\Omega}_S^{i+1}(N) \to \cdots \to \overline{\Omega}_S^r((r-1)N)/\overline{\Omega}_S^r((r-i-1)N) \to 0$$

is a resolution of $\overline{\Omega}_A^i$.

Taking associated sheaves on $X = \mathrm{Proj}(A)$, we obtain the resolution of the sheaf $\underline{\overline{\Omega}}_A^i$:

$$0 \to \underline{\overline{\Omega}}_A^i \to \overline{\Omega}_{\mathbb{P}}^i(N)/\overline{\Omega}_{\mathbb{P}}^i \to \cdots \to \overline{\Omega}_{\mathbb{P}}^r((r-1)N)/\overline{\Omega}_{\mathbb{P}}^r((r-i-1)N) \to 0 .$$

4.1.7. We are almost at the goal. It remains to show that the sheaf $\underline{\overline{\Omega}}_A^i$ coincides with the sheaf $\tilde{\Omega}_X^i$ defined as in 2.2.4 by setting $\tilde{\Omega}_X^i = j_*(\Omega_U^i)$, where $U = X - \mathrm{Sing}(X)$.

Let Z be an open set of nonsingular points of X such that $\tilde{\Omega}_X^i|Z = \Omega_Z^i$ and $\tilde{\Omega}_{\mathbb{P}}^i \otimes_{\mathcal{O}_{\mathbb{P}}} \mathcal{O}_Z = \Omega_{\mathbb{P}}^i \otimes_{\mathcal{O}_{\mathbb{P}}} \mathcal{O}_Z$. We have the exact sequence of locally free sheaves

$$0 \to \mathcal{N}_{X/\mathbb{P}} | Z \xrightarrow{d} \Omega_{\mathbb{P}}^1 \otimes_{\mathcal{O}_{\mathbb{P}}} \mathcal{O}_Z \to \Omega_Z^1 \to 0$$

where $\mathcal{N}_{X/\mathbb{P}} | Z = \mathcal{O}_Z(-N)$ is the normal sheaf of $Z \to \mathbb{P}$.

This sequence determines exact sequences

$$0 \to \Omega_Z^i \to \Omega_{\mathbb{P}}^i(N) \otimes_{\mathcal{O}_{\mathbb{P}}} \mathcal{O}_Z \xrightarrow{d} \Omega_Z^{i+1}(N)$$

which can be extended to the right to obtain the resolution

$$0 \to \Omega_Z^i \to \Omega_{\mathbb{P}}^i(N) \otimes_{\mathcal{O}_{\mathbb{P}}} \mathcal{O}_Z \to \Omega_{\mathbb{P}}^{i+1}(2N) \otimes_{\mathcal{O}_{\mathbb{P}}} \mathcal{O}_Z \to \cdots .$$

Since

$$\Omega_{\mathbb{P}}^{i+k}((1+k)N) \otimes_{\mathcal{O}_{\mathbb{P}}} \mathcal{O}_Z \cong \Omega_{\mathbb{P}}^{i+k}((1+k)N)/\Omega_{\mathbb{P}}^{i+k}(kN)| Z$$

we see that this resolution is the resolution of $\underline{\overline{\Omega}}_A^i$ (4.1.6) restricted on Z. Hence

$$\underline{\overline{\Omega}}_A^i | Z = \Omega_Z^i$$

and we obtain that

$$\overline{\Omega}^i_A = \widetilde{\Omega}^i_X .$$

Thus, we have constructed the resolution of $\widetilde{\Omega}^i_X$

$$0 \to \widetilde{\Omega}^i_X \to \Omega^i_{\mathbb{P}}(N)/\Omega^i_{\mathbb{P}} \to \cdots \to \Omega^r_{\mathbb{P}}((r-i+1)N)/\Omega^r_{\mathbb{P}}((r-1)N) \to 0 .$$

4.2. The Griffiths theorem.

This theorem generalizes for weighted hypersurfaces a result of [11] and allows to calculate the cohomology $H^i(X,\widetilde{\Omega}^i_X)$ as certain quotient spaces of differential forms on $\mathbb{P}$ with poles on X.

4.2.1. Denote by K^p the pth component of the resolution of $\widetilde{\Omega}^i_X$ from 4.1.7:

$$K^p = \overline{\Omega}^{i+p}_{\mathbb{P}}((p+1)N)/\overline{\Omega}^{i+p}_{\mathbb{P}}(pN) .$$

Using the exact sequence

$$0 \to \overline{\Omega}^{i+p}_{\mathbb{P}}(pN) \to \overline{\Omega}^{i+p}_{\mathbb{P}}((p+1)N) \to K^p \to 0$$

and the theorem of Bott-Steenbrink (2.3.4) we obtain that

$$H^q(X,K^p) = H^q(\mathbb{P},K^p) = 0 , \quad q>0 , \quad p>0$$

$$H^q(X,K^0) = H^q(\mathbb{P},K^0) = H^{q+1}(\mathbb{P},\overline{\Omega}^i_{\mathbb{P}}) = \begin{cases} k, & q=i-1 \\ 0, & q\neq i-1 \end{cases} .$$

Put

$$L = \mathrm{Ker}(K^1 \to K^2) .$$

Then we have the exact sequence of sheaves

$$0 \to \widetilde{\Omega}^i_X \to K^0 \to L \to 0$$

which gives the exact cohomology sequence

$$\cdots \to H^{q-1}(X,L) \to H^q(X,\widetilde{\Omega}^i_X) \to H^q(X,K^0) \to H^q(X,L) \to \cdots .$$

The sequence

$$0 \to L \to K^1 \to K^2 \to \cdots \to K^{r-i-1} \to 0$$

is an acyclic resolution of L. Thus we have

$$H^q(X,L) = H^q(\Gamma(X,K^\cdot)) = 0 , \quad q>r-i-2$$

$$H^{r-i-2}(X,L) = \Gamma(X,K^{r-i-1})/\mathrm{Im}\Gamma(X,K^{r-i-2}) =$$

$$= \Gamma(\mathbb{P},\overline{\Omega}^r_{\mathbb{P}}((r-i)N)/\Gamma(\mathbb{P},\overline{\Omega}^r_{\mathbb{P}}((r-i-1)N)) + \mathrm{Im}\Gamma(\overline{\mathbb{P}},\overline{\Omega}^{r-1}_{\mathbb{P}}((r-i-1)N) .$$

4.2.2. Theorem (Weak Lefschetz theorem). The homomorphism

$$H^q(X,\widetilde{\Omega}^i_X) \to H^q(X,K^0) \simeq H^{q+1}(\mathbb{P},\Omega^i_{\mathbb{P}})$$

is an isomorphism, if $q>r-i-1$ and an epimorphism if $q=r-i-1$.

Proof. Follows from the exact cohomology sequence 4.2.1 and the above calculation of $H^q(X,L)$.

Corollary. Assume that $k = \mathbb{C}$. Then we have an isomorphism of the Hodge structures

$$H^n(X,\mathbb{C}) \simeq H^{n+1}(\mathbb{P},\mathbb{C})$$

if $n \neq r-1$ and an epimorphism

$$H^{r-1}(X,\mathbb{C}) \to H^r(\mathbb{P},\mathbb{C}) .$$

For $n \geq r-1$ this directly follows from the theorem. For $n < r-1$ we use the Poincare duality for V-varieties (which are rational homology varieties).

Since the Hodge structure of $\mathbb{P}$ is known and very simple (2.3.6) we see that, as in the classic case, only cohomology $H^{r-1}(X)$ are interesting.

4.2.3. Put

$$h_0^{i,r-i-1}(X) = h^{i,r-i-1}(X) - a$$

where

$$a = \begin{cases} 1, & r = 2i \\ 0, & r \neq 2i . \end{cases}$$

Then we obtain that

$$h_0^{i,r-i-1}(X) = \dim_k H^{r-i-2}(X,L) .$$

Hence by calculations of 4.2.1 we obtain

Theorem (Griffiths-Steenbrink).

$$h_0^{i,r-i-1}(X) = \dim_k(\Gamma(\mathbb{P},\Omega^r_{\mathbb{P}}((r-i)N))/\Gamma(\mathbb{P},\Omega^r_{\mathbb{P}}((r-i-1)N)) + \operatorname{Im}\Gamma(\mathbb{P}, \Omega^{r-1}_{\mathbb{P}}((r-i-1)N))) .$$

4.3. Explicit calculation.

4.3.1. Let

$$\theta_f = \left(\frac{\partial f}{\partial T_0},\dots,\frac{\partial f}{\partial T_r} \right)$$

be the jacobian ideal with respect to a generator $f \in S(Q)_N$ of the ideal of the affine quasicone C_X of a weighted quasismooth hypersurface $X \subset \mathbb{P}(Q)$.

By the Euler formula 2.1.2 each $\frac{\partial f}{\partial T_i}$ is a homogeneous element of $S(Q)$ of degree $N - q_i$. Hence the ideal θ_f is homogeneous and the quotient space $S(Q)/\theta_f$ has a natural gradation. Since θ_f is m_0-primary this quotient space is finite dimensional.

4.3.2. Theorem (Steenbrink). Assume that $\mathrm{char}(k) = 0$. Then

$$h_0^{i,r-i-1}(X) = \dim_k (S(Q)/\theta_f)_{(r-i)N-|Q|} \, .$$

Proof. We have a natural isomorphism of graded $S(Q)$-modules:

$$\Omega_S^{r+1}/\Omega_S^r(-N)\wedge df = (S(Q)/\theta_f)(-|Q|) \, .$$

Since (see the proof of theorem 2.3.2)

$$\Gamma(\mathbb{P},\overline{\Omega}_{\mathbb{P}}^{i}(a)) = (\overline{\Omega}_S^{i})_a \, , \qquad \forall a \in \mathbb{Z}$$

and the differential $\Gamma(\mathbb{P},\Omega_{\mathbb{P}}^{i}(a)) \to \Gamma(\mathbb{P},\Omega_{\mathbb{P}}^{r+1}(a+N))$ corresponds to the operator d_f from 4.1.1, we can reformulate theorem 4.2.3 in the following form:

$$h_0^{i,r-i-1}(X) = \dim_k (\overline{\Omega}_S^{r}/d_f\overline{\Omega}_S^{r-1}(-N) + f\overline{\Omega}_S^{r}(-N))_{(r-i)N} \, .$$

Thus it remains to construct an isomorphism of graded S-modules

$$\overline{\Omega}_S^{r}/d_f\overline{\Omega}_S^{r-1}(-N) + f\overline{\Omega}_S^{r}(-N) \cong \Omega_S^{r+1}/\Omega_S^{r}(-N)\wedge df \, .$$

By property (iii) of lemma 2.1.3, we obtain that the k-linear map

$$d : \overline{\Omega}_S^{r} \to \Omega_S^{r+1}$$

is in fact an isomorphism of graded S-modules (here we use that $\mathrm{char}(k) = 0$!). By property (iii) of lemma 4.1.1, we set

$$d(d_f\overline{\Omega}_S^{r-1}(-N)) \subset \Omega_S^{r}(-N)\wedge df \, .$$

In fact, we have here an equality. Since d is S-linear it is sufficient to show that all forms

$$dx_{i_1}\wedge\ldots\wedge dx_{i_r}\wedge df \in d(d_g\overline{\Omega}_S^{r-1}(-N))) \, .$$

But

$$d(d_f(\Delta(dx_{i_1}\ldots dx_{i_r}))) = d(d_f(\sum_{s=1}^{r-1}(-1)^{s+1}x_{i_s}\,dx_{i_1}\wedge\ldots\wedge \widehat{dx_{i_s}}\wedge\ldots\wedge dx_{i_r})) =$$

$$= d(rf dx_{i_1}\wedge\ldots\wedge dx_{i_r} + (-1)^r a(\Sigma(-1)^{s+1}x_{i_s}\,dx_{i_1}\wedge\ldots\wedge\widehat{dx_{i_s}}\wedge\ldots\wedge dx_{i_r})\wedge df) =$$

$$= c\; dx_{i_1}\wedge\ldots\wedge dx_{i_r}\wedge df$$

where a,c are some rational numbers which we are too lazy to write down explicitly.

Thus d induces an isomorphism

$$\overline{\Omega}_S^{r}/d_f\overline{\Omega}_S^{r-1}(-N) \cong \Omega_S^{r+1}/\Omega_S^{r}(-N)\, df \, .$$

We have

$$d(f\Omega_S^{r}) \subset df\wedge\Omega_S^{r} + f\Omega_S^{r+1} \, .$$

But, since $f \in \theta_f$ (the Euler formula), $f\Omega_S^{r+1} \subset \Omega_S^{r}\wedge df$ and hence

$$d(f\Omega_S^{r}) \subset df\wedge\Omega_S^{r} \, .$$

Thus

$$f\overline{\Omega}_S^r(-N) \subset d_f\Omega_S^{r-1}(-N) ,$$

and we are through.

4.3.3. The theorem above can be reformulated in the following form. Define a function $\ell : \mathbb{Z}_{\geq 0}^{r+1} \to \mathbb{Z}$ by

$$\ell(a) = \frac{1}{N} \sum_{i=0}^{r} (a_i + 1)q_i , \quad a = (a_0,\ldots,a_r) \in \mathbb{Z}_{\geq 0}^{r+1} .$$

Let $\{T^a\}_{a\in J}$ be a set of monomials of $S(Q)$ whose residues mod θ_f generate the basis of the space $S(Q)/\theta_f$ (such monomials are called basic monomials). Then

$$h_0^{i,r-i-1}(X) = \#\{a \in J : \ell(a) = r-i\} .$$

4.4. Examples and supplements.

4.4.1. Suppose $f(T_0,\ldots,T_r) \in S$ is of the form

$$f(T_0,\ldots,T_r) = T_r^N + g(T_0,\ldots,T_{r-1}) .$$

If $(T^b)_{b\in J}$, are basic nomomials for $g(T_0,\ldots,T_{r-1})$ considered as elements of $S(q_0,\ldots,q_{r-1})$, then the set $\{T^bT_r^{b_r} : b \in J' , 0 \leq b_r \leq N-2\}$ is the set of basic monomials for f.

This implies that

$$h_0^{i,r-i-1}(X) = \#\{b \in J' : \ell(b) + \frac{b_r+1}{N} = r - i\}$$
$$= \#\{b \in J' : r - i - 1 < \ell(b) < r - i\} .$$

This formula was obtained in [11] in the homogeneous case.

4.4.2. More generally, if

$$f(T_0,\ldots,T_r) = g(T_0,\ldots,T_{r-1}) + T_r^m$$

then

$$h_0^{i,r-i-1}(X) = \#\{b \in J' : r - i - 1 + \frac{m}{N} \leq \ell(b) \leq r - i - \frac{m}{N}\} .$$

For example, if g is homogeneous then X is a multiple space (3.5.4) and we obtain

$$h_0^{i,r-1-i}(X) = \#\{b \in \mathbb{Z}^r : (r-i-2)N + m \leq |b| \leq (r-i-1)N-m , \quad 0 \leq b_j \leq N-2\}$$

where $|b| = b_0 + \cdots + b_{r-1}$ (cf. III, 8.8).

This can be written in more explicit form

$$h_0^{i,r-i-1} = \sum_{s=(r-i-2)N-m}^{(r-i-1)N-m} c_s$$

where c_s is the coefficient at t^s in $(1 + \cdots + t^{N-2})^r$.

4.4.3. Let $Y \subset \mathbb{P}(q_0,\dots,q_{r-1}) = \mathbb{P}(Q')$ be the hypersurface defined by the polynomial $g(T_0,\dots,T_r)$ from 4.4.1. Then

$$h_0^{i,r-i-2}(Y) = \#\{b \in J' : \ell(b) = r-i-1\} .$$

Assume now that $k = \mathbb{C}$. The exact sequence of Hodge structures

$$\cdots \to H^i(X) \to H^i(X-Y) \to H^{i-1}(Y)(-1) \to H^{i+1}(X) \to \cdots$$

(dual to the compact cohomology sequence) determines the morphisms of the Hodge structures:

$$i_* : H^{i-1}(Y)(-1) \to H^{i+1}(X)$$

which are obviously induced by the analogous morphisms

$$H^i(\mathbb{P}(Q'))(-1) \to H^{i+2}(\mathbb{P}(Q)) .$$

Applying the Weak Lefschetz theorem (4.2.2) we have that i_* is an isomorphism if $i \neq 0$, $r-1$. Thus for $U = X - Y$

$$H^i(U) = 0 , \quad i \neq 0,r-1 .$$

The Hodge structure on $H^{r-1}(U)$ has the following form

$$Gr_i^W(H^{r-1}(U)) = 0 , \quad i \neq r,r-1$$
$$Gr_r^W(H^{r-1}(U)) = H^{r-2}(Y)(-1)_0$$
$$Gr_{r-1}^W(H^{r-1}(U)) = H^{r-1}(H)_0 ,$$

where

$$H^{r-1}(X)_0 = \mathrm{Coker}(H^{r-3}(Y)(-1) \to H^{r-1}(X))$$
$$H^{r-2}(Y)_0 = \mathrm{Ker}(H^{r-2}(Y)(-1) \longrightarrow H^r(X)) .$$

For the Hodge numbers $h^{p,q}(U)$ we obtain (cf. [19])

$$h^{p,q}(U) = 0 , \quad \text{if } p+q \neq r-1,r$$
$$h^{i,r-i-1}(U) = h_0^{i,r-i-1}(X) = \#\{b \in J' : r-i-1 < \ell(b) < r-i\}$$
$$h_0^{i,r-i}(U) = h_0^{i-1,r-i-1}(Y) = \#\{b \in J' : \ell(b) = r-i-1\}$$

where we recall that

$$J' = \{b \in \mathbb{Z}^r_{\geq 0} : T^b \text{ are basic monomials for } g(T_0,\dots,T_{r-1})\} .$$

4.4.4. The calculations of 4.4.3 presents an interest since the open affine subset $U \subset X$ is isomorphic to the nonsingular affine variety in $\mathbb{A}^r$ with the equation

$$g(x_0,\dots,x_{r-1}) = 1 .$$

This variety plays an important part in the theory of critical points of analytic functions. The cohomology space $H^{r-1}(U)$ is isomorphic to the space of he vanishing cohomology of the isolated critical point $0 \in C^r$ of the analytic unction $t = g(z_0,\dots,z_{r-1})$ ([18]). Its dimension (the Milnor number)

$$\mu = \dim_C C[T_0,\dots,T_{r-1}]/\theta_g = \#J' .$$

t can be seen from above as follows:

$$\dim_C H^{r-1}(U) = \sum_{i=0}^{r-1} h_0^{i,r-i-1}(X) + \sum_{i=1}^{r-2} h_0^{i-1,r-i-1}(Y) =$$

$$= \#\{b \in J' : \ell(b) < r\}$$

nd so we have to show that for any basic monomial T^b $\ell(b) < r$ or, equivalently,

$$\deg(T^b) < \sum_{i=0}^{r-1} (N-q_i) = rN - |Q'| .$$

et

$$\mu_k = \#\{b \in J' : \deg(T^b) = k\}$$

$$\chi_g(z) = \sum_k \mu_k z^k .$$

hen ([1])

$$\chi_g(z) = \prod_{i=0}^{r-1} \frac{z^{N-q_i}-1}{z^{q_i}-1} .$$

t is clear that $\chi_g(z)$ is of degree $n = rN - 2|Q'|$ and hence for $k > n$ $\mu_k = 0$. his proves the assertion above.

Note that the Hodge numbers of $H^{r-1}(U)$ can be expressed in terms of μ_k as ollows

$$h^{i,r-i-1}(U) = \Sigma_{(r-i-1)N-|Q| < k < (r-i)N-|Q|} \mu_k$$

$$h^{i,r-i}(U) = \mu_{(r-i-1)N-|Q|} .$$

he symmetry of the Hodge numbers is in the accord with the symmetry of μ_k:

$$\mu_k = \mu_{n-i} .$$

.4.5. Let $X = V_n(Q)$ be a quasismooth surface (r=3). We know that

$$h^{0,2}(X) = h^{2,0}(X) = \#\{a \in J : \ell(a) = 1\} = \mu_{N-|Q|}$$

$$h^{1,1}(X) = \#\{a \in J : \ell(a) = 2\} = \mu_{2N-|Q|}$$

$$b_2(X) = 2h^{2,0}(X) + h^{1,1}(X) = 2\mu_{N-|Q|} + \mu_{2N-|Q|} ,$$

here

$$\Sigma \mu_k z^k = \prod_{i=0}^{r} (z^{N-q_i}-1)/(z^{q_i}-1) .$$

t is clear that $\mu_{N-|Q|} = a_{N-|Q|}$ in notations of 3.4.4. In case $\mathbb{P}(Q) = \mathbb{P}^3$ we have

$$\Sigma\mu_k z^k = ((z^{N-1}-1)/(z-1))^4 = (1+z+\cdots+z^{N-2})^4 .$$

REFERENCES

[1] V.I. ARNOL'D: Normal forms of functions, Uspehi Mat. Nauk, 29, No.2 (1974), 11-49 (in Russian; Engl. Transl.: Russian Math. Surveys, 29, No. 2 (1974), 10-50).

[2] M. ATIYAH, I. MACDONALD: Introduction to commutative algebra, Addison-Wesley Publ. Comp., 1969.

[3] N. BOURBAKI: Groupes et Algebres de Lie, ch. IV-VI, Hermann, 1968.

[4] F. CATANESE: Surfaces with $k^2 = p_g = 1$ and their period mapping; Proc. Copenhagen Sum. Meeting on Alg. Geom., Lect. Notes in Math. vol. 732, 1-29, Springer-Verlag, 1979.

[5] V. DANILOV: Geometry of torical varieties. Uspehi Mat. Nauk 33, No. 2 (1978) 83-134 (in Russian; Engl. Transl.: Russian Math. Surveys, 33, No. 2 (1978), 97-154).

[6] V. DANILOV: Newton polytopes and vanishing cohomology. Funkt. Anal. Priloz 13, No. 2 (1979), 32-47 (in Russian; Engl. Transl. Funct. Anal. Appl. 13, No. 2 (1979), 103-114).

[7] C. DELORME: Espaces projectifs anisotropes, Bull. Soc. Math., France, 103, 1975, 203-223.

[8] I. DOLGACHEV: Automorphic forms and quasihomogeneous singularities (in preparation).

[9] I. DOLGACHEV: Newton polyhedra and factorial rings, J. Pure and App. Algebra 18 (1980), 253-258; 21 (1981), 9-10.

[10] R. FOSSUM: The divisor class group of a Krull domain, Ergeb. Math. Bd. 74, Springer-Verlag. 1973.

[11] P. GRIFFITHS: On the periods of certain rational integrals I, Ann. Math. 90 (1969), 460-495.

[12] A. GROTHENDIECK: Elements de Geometrie Algebrique, ch. 2, Publ. Math. de l'IHES, No. 8, 1961.

[13] A. GROTHENDIECK: Cohomologie locale des faisceaux cohérents et théoremes de Lefschetz locaux et globaux (SGA2), North Holland Publ., 1968.

[14] H. HAMM: Die Topologie isolierten Singularitäten von vollstandigen Dürchschnitten koplexer Hyperflächer Dissertation. Bonn. 1969.

[15] A. HOVANSKIĬ: Newton polyhedra and the genus of complete intersections, Funkt. Anal.: Priloz. 12, No 1 (1978), 51-61 (in Russian; Engl. Transl.: Funct. Anal. Appl., 12, No 1 (1978), 38-46).

[16] E. KUNZ: Holomorphen Differentialformen auf algebraischen Varietäten mit Singularitäten. Manuscripta Math., 15 (1975), 91-108.

[17] YU. I. MANIN: Cubic forms, Moscow, 1972; (in Russian: Engl. Transl. by North Holland, 1974).

[18] J. MILNOR: Singular points of complex hypersurfaces, Ann. Math. Studies, No 61, Princeton, 1968.

[19] S. MORI: On a generalization of complete intersections, Journ. Math. Kyoto Univ., vol. 15, 3 (1975), 619-646.

[20] P. ORLIK, P. WAGREICH: Isolated singularities with $\mathbb{C}^*$-action, Ann. Math., 93 (1971), 205-228.

[21] H. PINKHAM: Singularités exceptionneles, la dualite étrange d'Arnold et les surfaces K3, C.R. Acad. Sci. Paris, 284 (1977), 615-617.

[22] H. POPP: Fundamentalgruppen algebraischer Mannigfaltigkeiten, Lect. Notes Math., vol. 176, Springer-Verlag 1970.

[23] J. STEENBRINK: Intersection forms for quasihomogeneous singularities, Comp. Math., v. 34, Fasc. 2, (1977), 211-223.

[24] A. TODOROV: Surfaces of general type with $K^2 = p_g = 1$, I. Ann. Scient. Ec. Norm. Sup., 13 (1980), 1-21.

[25] P. WAGREICH: Algebras of automorphic forms with few generators. Trans. AMS, 262 (1980), 367-389.

A PATHOLOGICAL EXAMPLE OF AN ACTION OF k^*

by

Jerzy Konarski

In this lecture we shall construct an example of the action of k^* on a normal variety in which there exists an orbit "starting" and "finishing" in the same connected component of the fixed point set. We shall also give some properties of k^*-actions, which will be needed in the following lecture.

1. Notations. Let k be an algebraically closed field of any characteristic. All varieties and morphisms will be defined over k. We shall consider algebraic actions of the multiplicative group k^* of the field k. We shall identify k^* with the open subset of the projective line P^1 consisting of the points different than 0 and ∞. Let us assume that there is given an action $\phi: k^* \times X \to X$ on a variety X. For each point $x \in X$, consider the morphism $\phi_x: k^* \to X$ defined by $\phi_x(t) = tx = \phi(t,x)$ for $t \in k^*$. If this morphism extends to a morphism $\bar{\phi}_x: k^* \cup \{0\} \to X$, we denote the value $\bar{\phi}_x(0)$ by $\lim_{t\to 0} tx$; similarly we define $\lim_{t\to\infty} tx$. If $F_1,\ldots,F_\ell$ denote the connected components of X^{k^*}, then the $X^+(F_i)$ give an invariant cover of X if X is complete, where $X^+(F) = \{x: \lim_{t\to 0} tx \in F\}$ (and similarly for the sets $X^-(F_i)$ where $X^-(F)$ is defined analogously as $\{x: \lim_{t\to\infty} tx \in F\}$). We call them the cells of the plus decomposition of X determined by the action of k^*, or simply the plus-cells of X. Analogously the minus-cells are defined.

The above decompositions were defined and studied by A. Bialynicki-Birula in [2] and [1] in the case, when the variety X was nonsingular. They have in this case very nice propoerties, see [2], [1]:

A). The cells $X^+(F_i)$ are locally closed in X and the projections along the orbits $\phi_i: X^+(F_i) \to F_i$, $\phi_i(x) = \lim_{t\to 0} tx$, $x \in X^+(F_i)$ are morphisms,

B). $\phi_i: X^+(F_i) \to F_i$ are algebraic bundles over F_i with affine spaces as fibres.

It is easy to construct non-normal surfaces with actions of k^* which do not satisfy the above conditions (e.g. identify suitable fixed points in the suitable linear action on the projective plane P^2), thus we shall restrict our assumptions.

2. Definition. An action of k^* on a variety X is called locally linear, if for every point $x \in X$ there exists a k^*-invariant open affine neighbourhood of x, which can be equivariantly embedded in an affine space with a linear action of k^*.

3. Theorem. (Sumihiro, [6]) If the variety X is normal, then every action of k^* on X is locally linear.

Let us consider a locally linear action of k^* on a complete variety X. Let U be an open affine invariant subset of X and let $j: U \to A^N$ be an equivariant embedding, as in the definition above. We may assume that the action of k^* on A^N is diagonal: $t(x_1,\dots,x_N) = (t^{m_1}x_1,\dots,t^{m_N}x_N)$ for every $t \in k^*$, $(x_1,\dots,x_N) \in A^N$ and for some weights $m_1,\dots,m_N$.

4. Remark. The set $X^+(X^{k^*} \cap U)$ is locally closed in X (closed in U) and the projection along the orbits is regular on $X^+(X^{k^*} \cap U)$. Indeed, $X^+(X^{k^*} \cap U) = U \cap j^{-1}(\{y \in A^N : \lim_{t\to 0} ty \text{ exists in } A^N\}) = U \cap j^{-1}(\{y = (y_1,\dots,y_N) \in A^N : \inf m_i < 0 \Rightarrow y_i = 0\})$. The projection is a morphism, since it is a restriction of the projection in A^N.

5. Lemma. The cells of the plus decomposition of X are constructible.

Proof. Let $U_1,\dots,U_j$ denote open affine subsets, as in definition 2, covering X. Then, for every $i = 1,\dots,\ell$, the cell $X^+(F_i)$ is constructible since it is a union of locally closed subsets $X^+(F_i \cap U_k)$, $k = 1,\dots,j$.

6. Remark. The lemma is true for arbitrary (not necessarily locally linear) actions. For the proof one uses normalization and the theorem of Sumihiro.

7. Corollary. There exists a cell which is a dense subset of X.

8. Definition. The dense cell is called the big cell. The connected component of the fixed point set contained in the big cell is called the source for the plus decomposition and the sink for the minus decomposition.

9. Theorem. Let F_1 denote the source of a locally linear action of k^* on a complete variety X. Then the big cell $X^+(F_1)$ is open in X, the component F_1 is irreducible and the projection $\phi_1: X^+(F_1) \to F_1$ is a morphism.

Proof. As above we choose a covering $X = \bigcup_{1\le k\le j} U_k$. There exists k, such that $X^+(F_1 \cap U_k)$ is open in X. We may assume $k = 1$. The set $X^+(F_1 \cap U_1)$ is irreducible as an open subset of an (irreducible) variety X, thus $F_1 \cap U_1$ is also irreducible as the image of $X^+(F_1 \cap U_1)$ under the regular map (Remark 4). Let F_1' denote the closure of $F_1 \cap U_1$ in X, then F_1' is an irreducible component of F_1. Let us fix a point $x \in F_1'$ and let us choose one of the open sets U_k, for example let it be U_2, containing x. The set $X^+(F_1' \cap U_1 \cap U_2)$ is open in $X^+(F_1' \cap U_2)$. On the other hand, it is open in $X^+(F_1' \cap U_1) = X^+(F_1 \cap U_1)$ and therefore it is open in X. Since $X^+(F_1' \cap U_2)$ is closed in U_2 (Remark 4) and contains the set

$X^+(F_1' \cap U_1 \cap U_2)$ open in X, it follows that $X^+(F_1' \cap U_2) = U_2$ (we use the irreducibility of X here). Thus we have found a neighbourhood $X^+(F_1' \cap U_2) = U_2$, $x \in U_2$, open in X and contained in $X^+(F_1')$. Since the point x was chosen arbitrarily, the set $X^+(F_1')$ is open in X. It follows also that $F_1' = F_1$: in fact, in the open neighbourhood $X^+(F_1')$ of F_1' there are no fixed points lying out of F_1'. At least the projection along the orbits is a morphism, since it is regular on each of the open subsets $X^+(F_1 \cap U_k)$, $k = 1,\dots,j$.

10. Remark. The irreducibility of the big cell is important in this proof. The same proof is good for any irreducible cell.

11. Remark. It follows from the theorem, that the source F_1 forms a cell of the minus decomposition.

Now we shall construct an example of an action of k^* on a normal variety (therefore the action will be locally linear) such that there exists a cell which is not locally closed and the projection along the orbits is not continuous on this cell. In this example, the essential property is the existence of an orbit for which both limits belong to the same component of the fixed point set.

12. Remark. Property A) doesn't follow from Remark 4.

13. Remark. Let $I_1,\dots,I_p$ denote all the irreducible components of the fixed point set. I don't know if the sets $X^+(I_j)$, $j = 1,\dots,p$, are locally closed and if the projections $X^+(I_j) \to I_j$ are regular. I missed it in my paper [5]. Also, I don't know if there may exist an orbit with both limits in one I_j.

Following Jurkiewicz [3], we shall consider k^*-actions on torus embeddings given by a homomorphism of k^* into the torus. First we recall the needed facts, details and proofs can be found in [4]. Let T denote an n-dimensional torus. A normal variety X with the given action of T is called a torus embedding, if X contains T as an open T-invariant subset and the action of T on X restricted to T is multiplication. The torus embedding X is called complete, if the variety X is complete, and is affine, if the variety X is affine.

Let M denote the group $X(T) = \mathrm{Hom}(T,k^*)$ of characters of the torus T, N the group $Y(T) = \mathrm{Hom}(k^*,T)$ of one-parameter subgroups of T, and let $M_{\mathbb{R}} = M \otimes_{\mathbb{Z}} \mathbb{R}$ and $N_{\mathbb{R}} = N \otimes_{\mathbb{Z}} \mathbb{R}$ be the corresponding vector spaces over the field of real numbers $\mathbb{R}$. The natural pairing $\langle\ ,\ \rangle$ extends linearly from $M \times N$ onto $M_{\mathbb{R}} \times N_{\mathbb{R}}$. A subset $\sigma \subset N_{\mathbb{R}}$ is called a rational convex polyhedral cone or simply a cone, if the following equivalent conditions are satisfied:

i) there exists a finite number of linear functionals $\ell_i, i=1,\dots,N$, defined over the field of rational numbers Q such that $\sigma = \{x\in N_{\mathbb{R}} : \ell_i(x) \geq 0 \text{ for } i=1,\dots,N\}$.

ii) there exists a finite number of vectors $x_i \in N_{\mathbb{R}}$, $i = 1,\dots,N$, defined over Q such that $\sigma = \{\sum_{i=1}^{N} \lambda_i x_i : \lambda_i \geq 0, i=1,\dots,N\}$.

For the given cone $\sigma \subset N_{\mathbb{R}}$ we denote by $\check{\sigma}$ its dual cone in $M_{\mathbb{R}}$, which consists of all the functionals $r \in M_{\mathbb{R}}$ satisfying $\langle r,a\rangle \geq 0$ for each $a \in \sigma$. If $\sigma = \{x : \ell_i(x) \geq 0, i=1,\dots,N\}$ is a cone in N_R, any subset $\sigma_1 = \sigma \cap \{x : \ell_i(x) = 0 \text{ for } i \in I\}$, where I is a subset in $\{1,\dots,N\}$, is called a face of σ. If σ_1 is a face of σ, we define the face $\sigma_1^{\perp}$ of the cone $\check{\sigma}$ as $\sigma_1^{\perp} = \{r \in \check{\sigma} : \langle r,\sigma_1\rangle = 0\}$. We define the interior of the face σ as $\text{Int}\,\sigma = \{w \in \sigma : \inf_{\chi\in\check{\sigma}-\sigma^{\perp}} \langle \chi,w\rangle > 0\}$.

Let a cone σ not contain any line (i.e. one-dimensional subspace) in $N_{\mathbb{R}}$. We denote by X_σ the corresponding torus embedding, i.e. $X_\sigma = \text{Spec}[k[\check{\sigma} \cap M]]$. There is a one-to-one correspondence between orbits of the action of T on X_σ and the faces of the cone σ. Namely, let σ_1 be a face of σ and let a be any point of $N \cap \text{Int}\,\sigma_1$. The point a is a one-dimensional subgroup of T, let us consider the orbit of the unit element $e \in T \subset X_\sigma$ under the action of a. Since $a \in \sigma$ and $\chi(a(t)) = t^{\langle\chi,a\rangle}$ for $\chi \in M$, $t \in k^*$, then $\langle\chi,a\rangle \geq 0$ for all $\chi \in \check{\sigma}$, and therefore the limit $e_a = \lim_{t\to 0} a(t)e$ exists in X_σ. The orbit 0^{σ_1} of the action of T on X_σ corresponding to the face σ_1 of the cone σ is defined as the orbit of the point e_a. This orbit may be described by the condition: $0^{\sigma_1} = \{p \in X_\sigma : \inf_{\chi\in\check{\sigma}\cap M} \chi(p) \neq 0 \Leftrightarrow \chi \in \sigma_1^{\perp}\}$ (note that $\chi(e_a) \neq 0 \Leftrightarrow \langle\chi,a\rangle = 0 \Leftrightarrow \chi(e_a) = 1$). The above correspondence has the following properties. Firstly, $\dim 0^{\sigma_1} + \dim \sigma_1 = \dim T$ ($\dim \sigma_1$ denotes here the dimension of the subspace spanned by σ_1). Secondly, σ_1 is a face of σ_2 if and only if $\overline{0^{\sigma_1}} \supset 0^{\sigma_2}$.

If $\Sigma = \{\sigma_\alpha\}$ is a rational partial polyhedral decomposition of $N_{\mathbb{R}}$, i.e. a finite set of cones satisfying the following conditions:

i) if σ is a face of σ_α, then $\sigma \in \Sigma$,

ii) for any α, β, $\sigma_\alpha \cap \sigma_\beta$ is a face of σ_α and σ_β,

then we can glue the X_{σ_α}'s together, obtaining a variety X_Σ. The torus embedding X_Σ is complete if $N_{\mathbb{R}} = \bigcup_{\sigma_\alpha\in\Sigma} \sigma_\alpha$. Given Σ and a one-parameter subgroup $a : k^* \to T$ of T, (i.e. $a \in N$), let us consider the induced action of k^* on X_Σ. We shall describe this action in terms of the rational partial polyhedral decomposition Σ.

<u>14. Lemma</u>. (Jurkiewicz) Let σ_1 be a face of σ, let p be any point in the orbit 0^{σ_1}, and let $a \in N$ (i.e. a is some one-parameter subgroup of T). Then the limit $p_a = \lim_{t\to 0} a(t)p$ belongs to the orbit 0^{σ_2} of the action of T on X_σ if and only if the face σ_2 satisfies the following condition: "for each

$w \in \mathrm{Int}\,\sigma_1$, there exists $\varepsilon > 0$ such that for any q, $0 < q < \varepsilon$, the point $w + qa$ belongs to $\mathrm{Int}\,\sigma_2$". We may rephrase this condition by saying that the vector a, when attached to the face σ_1, points into the face σ_2.

Proof. We use the equality $\chi(a(t)p) = t^{<\chi,a>}\chi(p)$ for $t \in k^*$, $\chi \in M$. The limit $p_a = \lim_{t\to 0} a(t)p$ exists in $X_\sigma = \mathrm{Spec}\, k[\check{\sigma} \cap M]$ if and only if for each character $\chi \in \check{\sigma} \cap M$, we have $\chi(p) = 0$ (i.e. $\chi \in \check{\sigma} \setminus \sigma_1^\perp$) or $<\chi,a> \geq 0$.

Assume that p_a exists in X_σ. Since $\chi(p_a) = \lim_{t\to 0} \chi(p)t^{<\chi,a>}$, then, for $\chi \in \check{\sigma} \cap M$, the following conditions are equivalent:

i) $\chi \in \check{\sigma} \setminus \sigma_2^\perp$

ii) $\chi(p_a) = 0$

iii) $\chi(p) = 0$ or $<\chi,a> > 0$

iv) $\chi \in \check{\sigma} \setminus \sigma_1^\perp$ or $<\chi,a> > 0$.

The last condition is equivalent to the condition from the assertion of the lemma: for each $w \in \mathrm{Int}\,\sigma_1$ and $\chi \in \check{\sigma} \setminus \sigma_2^\perp$, $<\chi, w+qa> = <\chi,w> + q<\chi,a> > 0$ for any q from some interval $(0,\varepsilon)$.

Now assume that p_a doesn't exist in X_σ, thus there exists a character $\chi \in \check{\sigma} \cap M$, such that $\chi(p) \neq 0$ (i.e. $<\chi,\sigma_1> = 0$) and $<\chi,a> < 0$. Then $<\chi, w+qa> < 0$, thus for any $q > 0$ the point $w+qa$ lies outside σ. Therefore, in this case there exist no faces in σ satisfying the condition of the lemma.

<u>15. Corollary</u>. The orbit 0^{σ_1} consists of fixed points of the action of k^* if and only if $a \in \mathrm{lin}\,\sigma_1$.

The above lemma is essential in the construction of our example. The idea is following: we want to find a normal variety (torus embedding) with an action of k^* such that there exists an orbit having both limits in the same connected component of the fixed point set. Let T be now a three-dimensional torus. Then $N_{\mathbb{R}}$ is a three-dimensional vector space. We shall describe the suitable rational partial polyhedral decomposition of $N_{\mathbb{R}}$ by drawing pictures - we avoid in this way writing down the coordinates of vectors spanning particular cones and the essence of the construction will be more clear.

Let us consider a rational partial polyhedral decomposition Γ as in diagram 1. In this picture is drawn the intersection of Γ with a sphere with a centre in $0 \in N_{\mathbb{R}}$ (strictly speaking, only some of the cones are drawn, over and below them may occur other ones). The corresponding torus embedding $X = X_\Gamma$ is drawn in the diagram 2. By the same letters we denote as well the faces in Γ as the corresponding subsets in X. The vector a is "vertical" and lies in linear hulls of "vertical" faces: three-dimensional $A,B,\ldots,U,Z,V,\ldots$ and two-dimensional $b,c,\ldots,u,v$. The corresponding points $A,B,\ldots,U,Z,V$ and connecting lines $b,c,\ldots,u,v$ form a connected component of the fixed point set X^{k^*} (Corollary 15).

Let us call this component Y. Note the position of the faces Z, z, w and α. There corresponds to them in X: the fixed point Z, curves z and w and the surface α. Points lying on the curves z and w and on the surface α are not fixed by k^*, their limits as $t \to 0$ are shown in diagram 2.

The cell $X^+(Y)$ of the plus decomposition of X is not locally closed: it contains the surface α, doesn't contain the curve z lying in the closure of α, but contains the point A lying on z. Also, projection along the orbits of the action of k^* is not continuous on $X^+(Y)$, because of the existence of the orbit w joining two fixed points A and V lying on Y. A rational partial polyhedral decomposition with required properties may be constructed for instance from seven three-dimensional cones (diagram 3).

The variety X_Γ obtained above is normal, singular, and complete but not projective. If it were nonsingular or projective and normal, it would have the property A) mentioned in the beginning. The latter case follows from the possibility of equivariant embedding of projective normal varieties into a projective space with a linear action of k^*, see [6].

REFERENCES

[1] A. Bialynicki-Birula, Some properties of the decompositions of algebraic varieties determined by actions of a torus, Bull. Acad. Polon. Sci. Ser. Sci. Math., Astr. et Phys. 24 (1976).

[2] A. Bialynicki-Birula, Some theorems on actions of algebraic groups, Annals of Math., 98 (1973), 480-497.

[3] J. Jurkiewicz, An example of an algebraic torus action which determines the nonfiltrable decomposition, Bull. Acad. Pol. Sci. Ser. Sci. Math., Astr. et Phys. 25 (1977), 1089-1092.

[4] G. Kempf, F. Knudsen, D. Mumford, B. Saint-Donat, Toroidal Embeddings I, Springer-Verlag, Lecture Notes in Mathematics, 339.

[5] J. Konarski, Decompositions of normal algebraic varieties determined by an action of a one-dimensional torus, Bull. Acad. Pol. Sci. Ser. Sci. Math., Astr. et Phys., 26 (1978), 295-300.

[6] H. Sumihiro, Equivariant completion, J. Math. Kyoto Univ., 14 (1974), 1-28.

Institute of Mathematics, Warsaw University

PKiN, 00-901 Warszawa

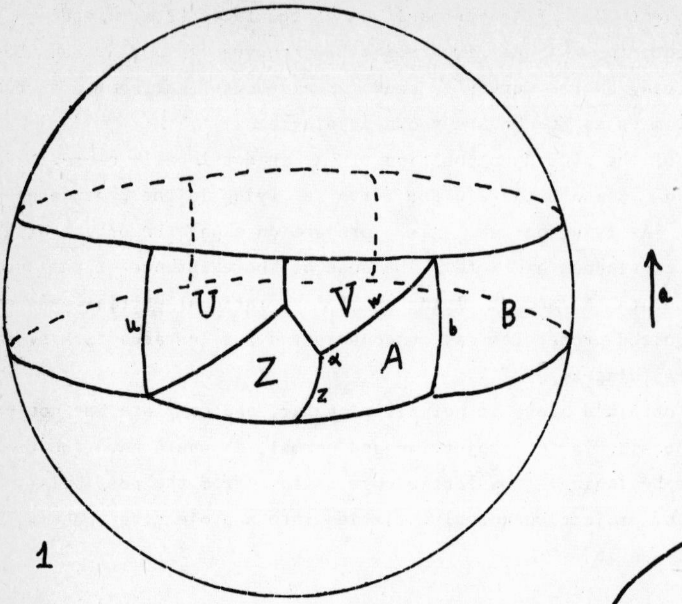

1

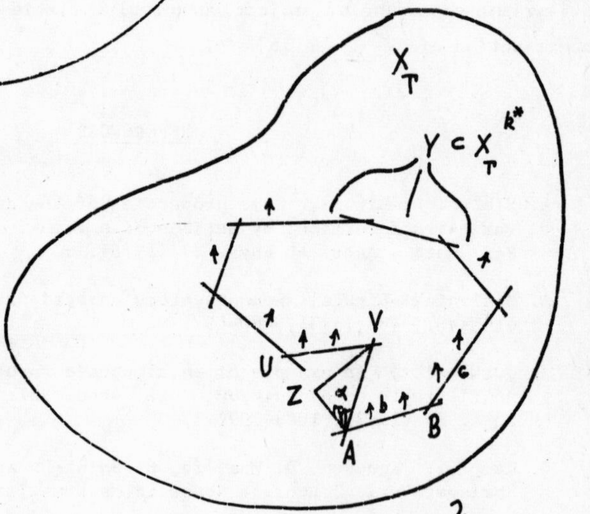

2.

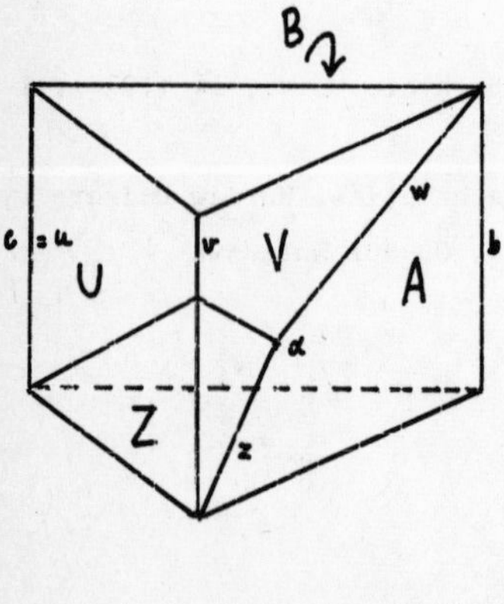

3.

Here, instead of a sphere, the intersection of Γ with the surface of a triangular prism is drawn (the prism has centre of mass at at $O \in N_{\mathbb{R}}$)

PROPERTIES OF PROJECTIVE ORBITS OF ACTIONS OF AFFINE ALGEBRAIC GROUPS*

by

Jerzy Konarski

Let a connected affine algebraic group act on a variety X. Then by X^* we will denote the set of all points $x \in X$ such that their orbits Gx are projective varieties. Our aim is to describe properties of the set X^*. The main results say that: a) the orbit type is locally constant on X^*; let us choose a connected component X_1^* of X^*, a parabolic subgroup P_1 representing the orbit type on X_1^* and a point $x \in X_1^*$, b) if X is normal and all orbits in X_1^* are G-isomorphic (e.g. if char k = 0), then X_1^* is G-isomorphic to a product $(X_1^*)^{P_1} \times Gx$, where $(X_1^*)^{P_1}$ denotes the set of points in X_1^* fixed by P_1. Then, assuming G to be reductive and X to be complete, we shall construct a decomposition of X into disjoint G-invariant subsets, each of them containing one of the connected components of X^*.

All varieties and morphisms, that will occur, will be defined over an algebraically closed field k. For an action $\sigma : G \times X \to X$ of an affine algebraic group G on a variety X and for any subset (or element) Y in X, GY will denote the image $\sigma(G \times Y)$. The set of fixed points of the action of G on X will be denoted by X^G. For any vector space V, $\mathbb{P}(V)$ will denote its projectivization. All varieties will be irreducible, unless otherwise stated. All algebraic groups will be affine. I will frequently use some facts on algebraic groups and parabolic and Borel subgroups. One can find them in [2]. The needed notions and properties of actions of the multiplicative group k^* may be found in the preceding lecture [4]. Several times I will use a theorem of Sumihiro, which says that if a connected algebraic group (respectively a torus) acts on a normal variety, we can cover this variety by open quasi-projective (respectively affine) invariant subsets; it is known also, that each of such subsets may be equivariantly embedded in a projective (resp. affine) space with a linear action of the group, see [5]. Theorems 2 and 11 were proved earlier by J.B. Carrell and A.J. Sommese for analytic actions of SL2 [3]. The construction of decomposition mentioned above follows A. Białynicki-Birula's construction for actions of SL2 [1].

1. Notations. We refer to the preceding lecture ([4]) for the basic notions and facts on actions of k^* used here without further comment, especially for the

---

*) This is a revised version of a preprint issued at Warsaw University in 1979.

notions of limits and of the plus and minus decompositions.

For any torus T (i.e. a product of a finite number of k^*'s), the character group $X(T)$ is a free abelian group whose dual group $Y(T)$ is the group of one-parameter subgroups of T. The duality $\langle\ ,\ \rangle : X(T)\times Y(T) \to \mathbb{Z}$ is given by $\langle\chi,\mu\rangle = m$ for $\chi \in X(T)$, $\mu \in Y(T)$ if and only if $\chi(\mu(t)) = t^m$ for all $t \in T$. For a finite number of nontrivial characters $\chi_1,\dots,\chi_n \in X(T)$, there exists an element $\mu \in Y(T)$ such that $\langle\chi_i,\mu\rangle \neq 0$ for $i = 1,\dots,n$. This follows, since for a finite number of hyperplanes in the vector space $Y(T)\otimes_{\mathbb{Z}}\mathbb{R}$ there exists in $Y(T)$ an element not lying on any one of them. Composition laws in $X(T)$ and $Y(T)$ are written additively.

If T is a maximal torus of an algebraic group G, then nontrivial weights of the adjoint representation of T on the tangent space T_eG at the identity $e \in G$ are called roots. If G is reductive, we have another definition: a character $\alpha \in X(T)$ is a root if there exists in G a one-parameter additive subgroup $g_\alpha : k^+ \to G$ such that $tg_\alpha(r)t^{-1} = g_\alpha(\alpha(t)r)$ for all $t \in T$, $r \in k^+$. The subgroup g_α is then unique up to a constant scalar. The set of roots of G with respect to T will be denoted by ϕ.

Let G be a connected algebraic group acting on an arbitrary variety X and let T be a fixed maximal torus in G. Then there exists such a one-parameter subgroup $\lambda : k^* \to T$ that

i) λ is regular, i.e. $\langle\alpha,\lambda\rangle \neq 0$ for all $\alpha \in \phi$,

ii) $X^{\lambda(k^*)} = X^T$.

This is so, because the action of T may be described with use of a finite number of nontrivial characters $\chi_i \in X(T)$, $i = 1,\dots,s$ (we normalize X and use the theorem of Sumihiro). Now we have to find an element $\lambda \in Y(T)$ such that $\langle\chi_i,\lambda\rangle \neq 0$ for $i = 1,\dots,s$ and $\langle\alpha,\lambda\rangle \neq 0$ for all $\alpha \in \phi$; this is possible as we have seen above. The one-parameter subgroup λ determines a Borel subgroup $B^+(\lambda)$ in G. Recall its definition. Let us choose any Borel subgroup B containing T. Let us consider the action of k^* on G/B induced by λ and by the action of T on G by conjugation (we shall call it the action of $\lambda(k^*)$). Then $B^+(\lambda)$ is the Borel subgroup corresponding to the source (consisting of one point, because λ is regular) of the action of $\lambda(k^*)$ on G/B. If G is reductive, there is another useful chracterization: $B^+(\lambda)$ is a subgroup in G, generated by T and by the subgroups g_α for all $\alpha \in \phi^+ = \{\alpha \in \phi : \langle\alpha,\lambda\rangle > 0\}$.

Note also that conversely, for a Borel subgroup $B \supset T$, there exists such a regular $\lambda \in Y(T)$ that $X^{\lambda(k^*)} = X^T$ and $B = B^+(\lambda)$. Really, $\{\lambda \in Y(T) : B = B^+(\lambda)\}$ is the intersection of the lattice $Y(T)$ with a nonempty open cone $\{y \in Y(T)\otimes_{\mathbb{Z}}\mathbb{R} : \langle\alpha,y\rangle > 0$ for $\alpha \in \phi^+\}$. If we add conditions $\langle\chi_i,y\rangle \neq 0$ for $i = 1,\dots,s$, we shall still obtain a nonempty set. Now we can formulate the first theorem.

<u>2. Theorem.</u> Let G be a reductive group acting on a variety X. Let T be

a fixed maximal torus in G and let $\lambda : k^* \to T$ be a regular one-parameter subgroup in T having the same fixed points in X as T. If x,y are points in X such that $\lim_{t\to 0} \lambda(t)x = y$, then $\lim_{t\to 0} \lambda(t)bx = y$ for all $b \in B^+(\lambda)$. In other words, $B^+(\lambda)$ preserves limits as $t \to 0$ of the action of $\lambda(k^*)$ on X. In particular, if X is complete, the cells of the plus decomposition of X determined by the action of $\lambda(k^*)$ are $B^+(\lambda)$-invariant.

3. Remark. The Borel subgroup $B^-(\lambda) = B^+(-\lambda)$, opposite to $B^+(\lambda)$, preserves limits as $t \to \infty$ of the action of $\lambda(k^*)$ on X.

Proof. Since $X^{\lambda(k^*)} = X^T$, T preserves limits of orbits of the action of $\lambda(k^*)$ on X: $\lim \lambda(t)sz = \lim s\lambda(t)z = s \lim \lambda(t)z = \lim \lambda(t)z$ for $s \in T$, $z \in X$, $t \in k^*$. Thus it suffices to show that $\lim_{t\to 0} \lambda(t)g_\alpha(r)x = y$ for $\alpha \in \phi^+$, $r \in k$. From the definition of g_α it follows that $\lambda(t)g_\alpha(r)x = g_\alpha(t^{<\alpha,\lambda>}r)\lambda(t)x$ for $t \in k^*$, $r \in k$. Denote $<\alpha,\lambda>$ by m (thus $m > 0$). Then $\lim_{t\to 0} \lambda(t)g_\alpha(r)x = \lim_{t\to 0} g_\alpha(t^m r)\lambda(t)x$.

Next, let us consider the following two morphisms, $f : k^* \to G$, given by $f(t) = g_\alpha(t^m r)$ for $t \in k^*$ and $g : k^* \to X$, $g(t) = \lambda(t)x$ for $t \in k^*$. They induce a morphism from k^* into the product $G \times X$ and, by composition with the action of G on X, a morphism $\rho : k^* \to X$ given by $\rho(t) = g_\alpha(t^m r)\lambda(t)x$ for $t \in k^*$. Since $m > 0$, f extends to a morphism $\overline{f} : k \to G$ with $\overline{f}(0) = g_\alpha(0) = e$, the identity. Also g extends to a morphism $\overline{g} : k \to X$, with $\overline{g}(0) = y$ (because $\lim_{t\to 0} \lambda(t)x = y$). Hence ρ extends to a morphism $\overline{\rho} : k \to X$ and $\lim_{t\to 0} g_\alpha(t^m r)\lambda(t)x = \overline{\rho}(0) = y$. This proves the theorem.

4. Corollary. Let G be a reductive group, T - a maximal torus in G and $\lambda : k^* \to T$ a regular one-parameter subgroup in T. Let $B = B^+(\lambda)$ and $X = G/B$. The action of G on X induced by multiplication on the left determines an action of $\lambda(k^*)$ on X. The plus decomposition of X determined by this action is then equal to the Bruhat decomposition.

Proof. Let $N(T)$ denote the normalizer of T in G and W - a Weyl group $N(T)/T$. The Bruhat cells are the cosets $BwB \subset G/B$ for $w \in W$. Since λ is regular, $X^{\lambda(k^*)} = X^T = \{gB \in X : T \subset gBg^{-1}\} = \{wB \in X : w \in W\}$ ([2]). For $w \in W$ let $X^+(w)$ denote the plus cell in X containing the fixed point wB. By the theorem $BwB \subset X^+(w)$. Since the cells $X^+(w)$, $w \in W$, are disjoint and the Bruhat cells BwB, $w \in W$, cover X, then for all $w \in W$, $BwB = X^+(w)$.

We shall apply the above theorem later. Now we shall pass to actions of arbitrary groups. Let us fix the following notation: G - a connected algebraic affine group acting on a variety X, T - a maximal torus in G and B - a Borel subgroup containing T. We shall denote by X^* the set of all points $x \in X$ such that their orbits Gx are projective varieties. Recall that the orbit of x is projective if and only if the isotropy subgroup G_x of x is parabolic, i.e. it contains some Borel subgroup. Recall also that an orbit type of an orbit Gx is the conjugacy class of the isotropy subgroup G_x in G.

5. Theorem. The orbit type is locally constant on X^*.

Proof. Let $\nu : X^\nu \to X$ be the normalization of X. The action of G lifts to an action on X^ν. Since the morphism ν is finite and parabolic subgroups are connected, for $y \in X^\nu$ and $x = \nu(y) \in X$, isotropy subgroups G_y and G_x are equal. Thus it is enough to prove the theorem for X normal. In this case we apply the theorem of Sumihiro and we see that it suffices to prove the theorem in the case when G acts linearly on a projective space $\mathbb{P}(V)$, i.e. there is given a homomorphism of algebraic groups $G \to \mathbb{P}GL(V)$; V denotes here a vector space. In general the homomorphism $G \to \mathbb{P}GL(V)$ doesn't lift to a homomorphism $G \to GL(V)$, in that case we replace G by the group $G \times_{\mathbb{P}GL(V)} GL(V)$. So, now we shall assume that the action on $\mathbb{P}(V)$ is induced by a rational representation of G in the vector space V. For $x \in \mathbb{P}(V)$, by $\hat{x}$ we shall denote any point in V lying over x.

If the orbit Gx of a point $x \in \mathbb{P}(V)$ is projective, the isotropy subgroup G_x contains some Borel subgroup B'. Since x is fixed with respect to B', there exists a character χ'_x of B' such that $b'\hat{x} = \chi'_x(b')\hat{x}$ for all $b' \in B'$. The subgroups B and B' are conjugate, i.e. $B = gB'g^{-1}$ for some $g \in G$, so χ'_x determines a character χ_x of B given by $\chi_x(gb'g^{-1}) = \chi'_x(b')$ for all $b' \in B'$. This definition does not depend on the choice of g by the theorem on a normalizer ([2], 11.15). Therefore we may define a function from $\mathbb{P}(V)^*$ to the group $X(B)$ of characters of $B : x \to \chi_x$ for $x \in \mathbb{P}(V)^*$.

6. Lemma. The above function is locally constant.

Proof. For a character $\chi \in X(B)$ let $V_\chi = \{y \in V : by = \chi(b)y\}$. The sets V_χ, $\chi \in X(B)$, are linear subspaces in V and the set $C = \{\chi \in X(B) : V_\chi \neq \{0\}\}$ is finite. We have $\mathbb{P}(V)^B = \bigcup_{\chi \in C} \mathbb{P}(V_\chi)$, the union of disjoint closed subvarieties. For $\chi \in C$ let $\phi_\chi : G/B \times \mathbb{P}(V_\chi) \to \mathbb{P}(V)^*$ be a morphism given by $\phi_\chi(gB,a) = ga$ for all $gB \in G/B$, $a \in \mathbb{P}(V_\chi)$; let Y_χ denote the image of ϕ_χ. All Y_χ's are closed in $\mathbb{P}(V)^*$, as images of projective varieties; they cover all $\mathbb{P}(V)^*$ and they are disjoint since each orbit in $\mathbb{P}(V)^*$ intersects $\mathbb{P}(V)^B$ in exactly one point. Hence, $\mathbb{P}(V)^*$ is a union of a finite number of disjoint closed subvarieties Y_χ such that on each of them the function $x \to \chi_x$ is constant.

We come back to the proof of the theorem. Let x be a point in $\mathbb{P}(V)^*$. For a suitable $g \in G$, $gx = y$ belongs to $\mathbb{P}(V)^B$. Denote by χ the character $\chi_x = \chi_y$ of B. We claim that the isotropy subgroup G_y is the largest subgroup containing B such that χ extends to it. We shall show that:

a) χ extends to G_y,

b) if χ extends to a subgroup $P \supset B$, then $p\hat{y} = \chi(p)\hat{y}$ for all $p \in P$.

(In other words, if χ extends on P, then $P \subset G_y$.)

a) is clear, since y is fixed by G_y. The proof of b) is following: the morphism $\zeta : P \to V$ given by $\zeta(p) = \chi(p)^{-1} \cdot p\hat{y}$ for all $p \in P$, is constant on each coset pB, $p \in P$, so it factors through P/B, whence it is constant, since P/B is a projective variety and V is affine.

This is what we wanted: since the function $x \to \chi_x$ is locally constant on $\mathbb{P}(V)^*$ and its values determine the orbit type, then the orbit type is also locally constant on $\mathbb{P}(V)^*$.

7. Corollary. ([1] Theorem 5) X^G is a union of some connected components of X^*.

Now, let us fix a connected component X_1^* in X^*. X_1^B will denote the component of X^B contained in X_1^*, i.e. $X_1^B = X^B \cap X_1^*$. Let a subgroup $P_1 \supset B$ represent the orbit type on X_1^*, i.e. P_1 is the isotropy subgroup of all points in X_1^B.

8. Theorem. If X is normal, then there exists a G-equivariant morphism $\pi : X_1^* \to X_1^B$.

Proof. As a map, π is determined uniquely. Since X is normal, we can cover it by open G-invariant quasi-projective subsets. For any set U of this covering, we have to show that the restriction $\pi|_{X_1^* \cap U} : X_1^* \cap U \to X_1^B \cap U$ is regular. We may embed U equivariantly in some projective space $\mathbb{P}$ with a linear action of G. Let $\mathbb{P}_1^*$ denote the component of $\mathbb{P}^*$ containing the image of $X_1^* \cap U$, and let $\mathbb{P}_1^B = \mathbb{P}_1 \cap \mathbb{P}^B$. It suffices to show that the G-equivariant projection $\pi_{\mathbb{P}} : \mathbb{P}_1^* \to \mathbb{P}_1^B$ is regular.

Note that since the radical $\mathrm{Rad}\, G$ acts trivially on $\mathbb{P}_1^*$, the action of G induces an action of a semisimple group $G' = G/\mathrm{Rad}\, G$ on $\mathbb{P}_1^*$ such that orbits of both actions are equal. Let T' be the image of T in G'; then T' is a maximal torus in G'. Let $\lambda' : k^* \to T'$ be such a one-parameter subgroup in T' that the Borel subgroup $B' = B/\mathrm{Rad}\, G$ in G' preserves limits $\lim_{t \to 0} \lambda'(t)x$ for $x \in \mathbb{P}_1^*$. We shall show that the action of $\lambda'(k^*)$ on $\mathbb{P}_1^*$ is locally linear (see [4] for a definition). Since $T \cap \mathrm{Rad}\, G$ is a direct summand in T, then λ' lifts to a one-parameter subgroup $\lambda : k^* \to T$. $\lambda(k^*) \subset G$, so it acts linearly on $\mathbb{P}$. Since the action of $\lambda'(k^*)$ on $\mathbb{P}_1^*$ is equal to the action of $\lambda(k^*)$, then it is locally linear. Now, let W denote the big (plus) cell of this action. Since $\mathbb{P}_1^* = G\mathbb{P}_1^B$ is irreducible, there exists a regular projection $(\lambda'(k^*)$-equivariant) $\pi' : W \to$ source of $\mathbb{P}_1^* = \mathbb{P}_1^{B^-}$, where B^- is the Borel subgroup opposite to B. Of course $\pi' = \pi_{\mathbb{P}|W}$, whence the projection $\pi_{\mathbb{P}}$ is regular on the neighbourhood W of $\mathbb{P}_1^{B-}$.

If we repeat the above construction, taking another Borel subgroup gBg^{-1} instead of B and composing the previously obtained projection with the translation $x \to gx$ for $x \in \mathbb{P}_1^*$, we obtain regularity of $\pi_{\mathbb{P}_*}$ on a neighbourhood of $g\mathbb{P}_1^{B^-}$. Since such neighbourhoods form a covering of $\mathbb{P}_1^*$, it follows that $\pi_{\mathbb{P}}$ is a morphism.

9. Before proving the next theorem, we recall, for the convenience of the reader, some facts about morphisms from [2]:

A) a dominant morphism $\phi : X \to Y$ is separable if and only if there exists a regular point $x \in X$ such that $\phi(x)$ is regular and the differential $d_x\phi$ is an epimorphism.

B) If $\phi : X \to Y$ is an open separable morphism, Y is normal and $Y = \phi(X)$, then ϕ is a quotient morphism, i.e. for any morphism $\psi : X \to Z$ constant on fibres of ϕ, there exists a unique morphism $\theta : Y \to Z$ such that $\psi = \theta \circ \phi$.

C) In particular, if ϕ is a bijective open separable morphism and its image is normal, then ϕ is an isomorphism.

D) If Y is a normal variety and $\phi : X \to Y$ is a dominant morphism such that all the irreducible components of the fibres of ϕ are of the same dimension, then ϕ is open.

E) For a vector space V, the natural projection $p : V \setminus \{0\} \to \mathbb{P}(V)$ and its restriction $p|_{p^{-1}(Z)} : p^{-1}(Z) \to Z$, where Z is any normal subvariety of $\mathbb{P}(V)$, are quotient morhpisms (by B and D).

10. Corollary (of Theorem 5). If $\operatorname{char} k = 0$, then all orbits in X_1^* are G-isomorphic to G/P_1. In other words, the orbits in X_1^* are rigid.

Proof. It follows from the separability of the morphism $\omega : G/P_1 \to Gx$, $\omega(gP_1) = gx$ for $g \in G$, $x \in X_1^B$ and from 9DC above.

Now we would like to mention some results obtained by Ewa Duma on deformations of two-dimensional orbits for actions of SL2 (unpublished). In SL2 we have the following types of one-dimensional subgroups (up to conjugacy) : T - a maximal torus; $N(T)$ - the normalizer of T; and N_m for m a natural number - the product of a unipotent subgroup by a cyclic group of order m. If there exists an action of SL2 such that the orbit type on an open subset U is H and in some point in the closure of U it is H', then we say H deforms to H'. The results are following: i) all three types deform to SL2 and to B, a Borel subgroup, ii) T deforms to $N(T)$ and to all N_m's, iii) $N(T)$ doesn't deform to T and it deforms to $N_m \Longleftrightarrow m$ is even, iv) N_m doesn't deform to T and to $N(T)$ and it deforms to $N_n \Leftrightarrow m$ divides n. There are also some results, still not complete, on finite isotropy subgroups.

We come back to projective orbits. Choose an arbitrary point y in the compoenent X_1^B. Let X_y^* denote the subset of X_1^* consisting of all points, the orbits of which are G-isomorphic to the orbit Gy. For instance, if $\operatorname{char} k = 0$, then $X_y^* = X_1^*$. Let $X_y^B = X_y^* \cap X^B$. There is an action of G on the product $Gy \times X_y^B$, induced by the action of G on Gy and by trivial action on X_y^B.

<u>11. Theorem</u>. If X is normal, then X_y^* is G-isomorphic to $Gy \times X_y^B$.

Proof. Let $\sigma : G/P_1 \times X_1^B \to X_1^*$ be a morphism induced by the action of G on X, i.e. $\sigma(gP_1, x) = gx$ for $gP_1 \in G/P_1$, $x \in X_1^B$. Let $\omega : G/P_1 \to Gy$ be a morphism given by $\omega(gP_1) = gy$ for $g \in G$. Since σ and ω are bijections, there exists a unique map τ such that the diagram commutes

$$\begin{array}{ccc} G/P_1 \times X_1^B & \xrightarrow{\sigma} & X_1^* \\ {\scriptstyle \omega \times id} \downarrow & \tau & \\ Gy \times X_1^B & & \end{array}$$

We have to prove that the sets X_y^B and X_y^* are varieties and that $\tau|_{Gy \times X_y^B} : Gy \times X_y^B \to X_y^*$ is an isomorphism. These properties are local and since X is normal, then by Sumihiro's theorem we may prove the statement for a linear action of G on a projective space. Thus, we shall assume that: there is given a linear action of G on a projective space $\mathbb{P}$; $\mathbb{P}_1^B$ denotes a component in $\mathbb{P}^B$ with a corresponding isotropy subgroup P_1; $\mathbb{P}_y^B = \{z \in \mathbb{P}_1^B : Gz$ is G-isomorphic to $Gy\}$; $\mathbb{P}_1^* = G \cdot \mathbb{P}_1^B$; $\mathbb{P}_y^* = G \cdot \mathbb{P}_y^B$. We have also a diagram, as before:

$$\begin{array}{ccc} G/P_1 \times \mathbb{P}_1^B & \xrightarrow{\sigma_{\mathbb{P}}} & \mathbb{P}_1^* \\ {\scriptstyle \omega \times id} \downarrow & \tau_{\mathbb{P}} & \\ Gy \times \mathbb{P}_1^B & & \end{array}$$

Now we have to show that $\mathbb{P}_y^B$ and $\mathbb{P}_y^*$ are subvarieties of $\mathbb{P}$ (i.e. are locally closed) and that $\tau_{\mathbb{P}}|_{Gy \times \mathbb{P}_y^B}$ is an isomorphism of $Gy \times \mathbb{P}_y^B$ onto $\mathbb{P}_y^*$. As in the proof of Theorem 5 we shall assume that the action of G on $\mathbb{P}$ is induced by a representation in a vector space V, i.e. $\mathbb{P} = \mathbb{P}(V)$, a projectivization of V, and the projection $p : V \setminus \{0\} \to \mathbb{P}(V)$ is G-equivariant. Denote $V \setminus \{0\}$ by A.

Let $z_0, \ldots, z_k \in \mathbb{P}_y^B$ be a maximal linearly independent subset, we may assume $z_0 = y$. For $i = 0, \ldots, k$ choose $\hat{z}_i \in A$ such that $p(\hat{z}_i) = z_i$. Now, let

$A_y = \mathrm{lin}(\hat{z}_0,\ldots,\hat{z}_k) \setminus \{0\} = p^{-1}(\mathbb{P}(z_0,\ldots,z_k)) \subset A$, where $\mathbb{P}(z_0,\ldots,z_k)$ means the projective subspace spanned by $z_0,\ldots,z_k$. By Theorem 5, for some character χ of B, we have $A_y \subset V_\chi = \{v \in V : bv = \chi(b)v$ for $b \in B\}$. The orbits $G\hat{z}_0,\ldots,G\hat{z}_k$ are G-isomorphic by the following lemma.

12. Lemma. Let $\hat{x}_1, \hat{x}_2 \in A$ be such points that $x_1 = p(\hat{x}_1)$, $x_2 = p(\hat{x}_2) \in P_1^B$. Then the orbits Gx_1 and Gx_2 are G-isomorphic if and only if the orbits $G\hat{x}_1$ and $G\hat{x}_2$ are G-isomorphic.

Proof. ($\Leftarrow$) Let $\alpha : G\hat{x}_1 \to G\hat{x}_2$ be a G-isomorphism. Then $p|_{G\hat{x}_1} : G\hat{x}_1 \to Gx_1$ and $p\circ\alpha : G\hat{x}_1 \to Gx_2$ are quotient morphisms by 9E. Since they have the same fibres, there exists a G-isomorphism $\alpha' : Gx_1 \to Gx_2$.

($\Rightarrow$) Let $\alpha' : Gx_1 \to Gx_2$ be a G-isomorphism. We shall construct an isomorphism of $G\hat{x}_1$ and $G\hat{x}_2$. Let us fix $i = 1,2$. Let U^- denote the unipotent part of B^-, then U^-B is open in G and since $\hat{x}_i$ is an eigenvector with respect to B, then $U^-k\hat{x}_i = U^-B\hat{x}_i$ is open in $G\hat{x}_i$ by 9D.

First we shall prove that $U^-k\hat{x}_i$ is isomorphic to $k \times U^-\hat{x}_i$. Let $\eta_i : k \times U^-\hat{x}_i \to U^-k\hat{x}_i$ be given by $\eta_i(a,z) = az$ for $a \in k$, $z \in U^-\hat{x}_i$. η_i is open by 9D and is bijective, since unipotent groups have no characters. So, by 9C it suffices to prove that the differential $d_{(1,x_i)}\eta_i$ is an isomorphism. Suppose the converse. As we see easily, the restrictions of $d_{(1,x_i)}\eta_i$ to the tangent spaces $T_{(1,x_i)}(k \times \{x_i\})$ and $T_{(1,x_i)}(\{1\} \times U^-\hat{x}_i)$ are isomorphisms, so we must have $T_{\hat{x}_i}k\hat{x}_i \subset T_{\hat{x}_i}U^-\hat{x}_i$. Let U' be a maximal unipotent subgroup in $GL(V)$ containing the image of U^-. Then we have $T_{\hat{x}_i}k\hat{x}_i \subset T_{\hat{x}_i}U'\hat{x}_i$. Since U' is an affine space and the action is linear, we may rewrite this inclusion in the form $k\hat{x}_i \subset U'\hat{x}_i$. This is a contradiction, because U' consists of unipotent elements and they have all eigenvalues equal one.

Next we note that $p|_{U^-\hat{x}_i} : U^-\hat{x}_i \to U^-x_i$ is an isomorphism. Indeed, it is bijective, open by 9D and separable, whence an isomorphism by 9C. Therefore the orbits $U^-\hat{x}_1$ and $U^-\hat{x}_2$ are isomorphic. It follows that in $G\hat{x}_1$ and $G\hat{x}_2$ there are isomorphic open subsets $U^-B\hat{x}_1$ and $U^-B\hat{x}_2$. Moreover, the isomorphism is compatible with the action, i.e. $g\hat{x}_1 \to g\hat{x}_2$ for $g \in U^-B$. We can now cover $G\hat{x}_i$, $i = 1,2$, by open subsets $gU^-B\hat{x}_i$ for $g \in G$. Note $gU^-B\hat{x}_i = gU^-g^{-1}gBg^{-1}g\hat{x}_i$ thus, repeating the above construction for the points $g\hat{x}_i$ and the Borel subgroup gBg^{-1}, we obtain an isomorphism of $gU^-B\hat{x}_1$ and $gU^-B\hat{x}_2$. Since all these isomorphisms are compatible with the action, we can glue them to an isomorphism of $G\hat{x}_1$ and $G\hat{x}_2$.

Continuing the proof of the theorem we shall show now that $\mathbb{P}_y^B$ is open in $\mathbb{P}(z_0,\ldots,z_k)$. Let $\beta_i : G\hat{z}_0 \to G\hat{z}_i$ be G-isomorphisms for $i = 0,\ldots,k$. Let $\beta : G\hat{z}_0 \times A_y \to A$ be a morphism given by $\beta(w,\hat{z}) = \beta(w, \sum_{i=0}^{k} \lambda_i\hat{z}_i) = \Sigma\lambda_i\beta_i(w)$ for

$w \in G\hat{z}_0$ and $\hat{z} = \sum_{i=0}^{k} \lambda_i z_i \in A_y$. For $\hat{z} \in A_y$ let $\beta_{\hat{z}} : G\hat{z}_0 \to G\hat{z}$ be given by $\beta_{\hat{z}}(w) = \beta(w,\hat{z})$ for $w \in G\hat{z}_0$. The morphisms β and $\beta_{\hat{z}}$ are G-equivariant. Since $A_y \subset V_\chi$, all points in A_y have the same isotropy subgroup (equal to $\ker\chi \subset P_1$). Thus, for all $\hat{z} \in A_y$, $\beta_{\hat{z}}$ is a bijective morphism onto a normal variety and it is open by 9D. Hence, $\beta_{\hat{z}}$ is an isomorphism if and only if its differential $d_{\hat{z}_0}\beta_{\hat{z}}$ is an isomorphism. Since $d_{\hat{z}_0}\beta_{\hat{z}} = \Sigma\lambda_i d_{\hat{z}_0}\beta_i$ and $d_{\hat{z}_0}\beta_i$ are isomorphisms, then $d_{\hat{z}_0}\beta_{\hat{z}}$ is an isomorphism if and only if the coefficients $\lambda_0,\dots,\lambda_k$ satisfy $\det(\Sigma\lambda_i d_{\hat{z}_0}\beta_i) \neq 0$. This condition gives a nonempty open subset in A_y. Thus $\{\hat{z}\in A_y : \beta_{\hat{z}}$ is an isomorphism$\}$ is open in A_y. It follows that $\mathbb{P}_y^B$ is open in $\mathbb{P}(z_0,\dots,z_k)$.

Let us now consider a morphism

$\pi = p|_{G\hat{z}_0} \times p|_{A_y} : G\hat{z}_0 \times A_y \to Gz_0 \times \overline{\mathbb{P}_y^B} = \mathbb{P}(z_0,\dots,z_k) \times Gz_0$. It is separable, open by 9D and its image is nonsingular. The morphism $p\circ\beta$ is constant on fibres of π, hence by the universal property 9B there exists a unique morphism $\overline{\tau}$ such that the following diagram commutes.

$$\begin{array}{ccccc} G\hat{z}_0 \times A_y & \xrightarrow{\beta} & G\,A_y & \subset & A \\ \pi\downarrow & & \downarrow & & \downarrow p \\ Gz_0 \times \overline{\mathbb{P}_y^B} & \xrightarrow{\overline{\tau}} & G\,\overline{\mathbb{P}_y^B} & \subset & \mathbb{P} \end{array}$$

We also see that $\mathbb{P}_y^*$ is locally closed in $\mathbb{P}$. Really, $\mathbb{P}_y^* = G\,\mathbb{P}_y^B$ is the image of the open set $Gz_0 \times \mathbb{P}_y^B$. Since $Gz_0 \times \mathbb{P}_y^B$ is the union of fibres of $\overline{\tau}$ and $\overline{\tau}$ is closed, $\mathbb{P}_y^*$ is open in $G\,\overline{\mathbb{P}_y^B} = \overline{\mathbb{P}_y^*}$.

13. Remark. If char k = 0, all the above part of the proof may be replaced by the sentence: the morphism $\omega : G/P_1 \to Gy$ is an isomorphism by Corollary 10, whence $\tau = \sigma\circ(\omega \times \mathrm{id})^{-1}$ is a morphism.

Now we want to prove that $\overline{\tau}$ is separable. For all $z \in \mathbb{P}_y^B$ we shall show that the differential $d_{(z_0,z)}\overline{\tau} : T_{(z_0,z)}Gz_0 \times \overline{\mathbb{P}_y^B} \to T_z\overline{\mathbb{P}_y^*}$ is a monomorphism. Since $\overline{\tau}|_{Gz_0 \times \{z\}} : Gz_0 \times \{z\} \to Gz$ and $\overline{\tau}|_{\{z_0\} \times \overline{\mathbb{P}_y^B}} : \{z_0\} \to \overline{\mathbb{P}_y^B} \to \overline{\mathbb{P}_y^B}$ are isomorphisms, it suffices to prove that $T_zGz \cap T_z\overline{\mathbb{P}_y^B} = \{0\}$. By Theorem 8 we have the projection $\pi : \overline{\mathbb{P}_y^*} \to \overline{\mathbb{P}_y^B}$. For $v \in T_zGz \cap T_z\overline{\mathbb{P}_y^B}$ we have $d_z\pi(v) = 0$, since $\pi|_{Gz} = \mathrm{const}$; and $d_z\pi(v) = v$, since $\pi|_{\overline{\mathbb{P}_y^B}} = \mathrm{id}$. Thus $v = 0$. This implies the separability of $\overline{\tau}$ since in $\mathbb{P}_y^B$ there are points which are regular on $\overline{\mathbb{P}_y^*}$, $d\overline{\tau}$ is then an isomorphism.

Since $\overline{\tau}$ is bijective and proper, it is a homeomorphism and since $Gz_0 \times \overline{\mathbb{P}_y^B}$ is normal, it factors through the normalization $\nu : (\overline{\mathbb{P}_y^*})^\nu \to \overline{\mathbb{P}_y^*}$ of the variety $\overline{\mathbb{P}_y^*}$, i.e. there exists a morphism $\overline{\tau}^\nu : Gz_0 \times \overline{\mathbb{P}_y^B} \to (\overline{\mathbb{P}_y^*})^\nu$ such that $\overline{\tau} = \nu\circ\overline{\tau}^\nu$. Since $\overline{\tau}$ is separable, so is $\overline{\tau}^\nu$ because they "coincide" on an open set. Since

$\overline{Gz_0 \times \mathbb{P}^B_y}$ is complete, $\overline{\tau}^\nu$ is a homeomorphism. Since its image is normal, then by 9C $\overline{\tau}^\nu$ is an isomorphism. It follows that $\overline{\tau}$ is finite (it is even a normalization). In particular $\tau_{\mathbb{P}} = \overline{\tau}|_{Gz_0 \times \mathbb{P}^B_y}$ is finite. As we showed above, for all $q \in Gz_0 \times \mathbb{P}^B_y$ $d_q\tau_{\mathbb{P}}$ is an isomorphism. Since $Gz_0 \times \mathbb{P}^B_y$ is nonsingular, then $\tau_{\mathbb{P}}$ is an isomorphism, thus the theorem is proved.

14. Corollary. If char $k = 0$, then X^*_1 is isomorphic to $G/P_1 \times X^B_1$. In analytic version this recently has been proved by M. Koras.

The following example illustrates the theorem and shows that if char $k \neq 0$, then

i) orbits passing through X^B_1 may not be G-isomorphic,

ii) the set X^*_1 may not be isomorphic (not only G-isomorphic !) to the product of an orbit with X^B_1,

iii) the set X^*_1 may be singular, even when X^B_1 is nonsingular.

15. Example. Let char $k = 2$ and let $G = SL2$ act on $X = k^1 \times \mathbb{P}^2$ by

$$\begin{pmatrix} a & b \\ c & d \end{pmatrix} \left(t, \begin{bmatrix} x_0 \\ x_1 \\ x_2 \end{bmatrix}\right) = \left(t, \begin{pmatrix} 1 & tac & tbd \\ 0 & a^2 & b^2 \\ 0 & c^2 & d^2 \end{pmatrix} \cdot \begin{bmatrix} x_0 \\ x_1 \\ x_2 \end{bmatrix}\right) \text{ for } \begin{pmatrix} a & b \\ c & d \end{pmatrix} \in G, \ t \in k^1,$$

$[x_0,x_1,x_2] \in \mathbb{P}^2$. Let $B = \{(\begin{smallmatrix} a & b \\ 0 & d \end{smallmatrix}) \in G\}$. Then X^B has two components: $X^B_1 = k^1 \times [0,1,0]$ and $X^B_2 = k^1 \times [1,0,0]$. The corresponding isotropy subgroups are B and G respectively. All orbits in X^*_1 are projective lines, but they are G-isomorphic only for $t \neq 0$. The set X^*_1 is singular and is not isomorphic to $\mathbb{P}^1 \times X^B_1$ which is nonsingular. On the other hand let $W = \{(t,z) \in X : t \neq 0\}$ and let H be an orbit in $X^*_1 \cap W$. Then $X^*_1 \cap W$ is G-isomorphic to $H \times (X^B_1 \cap W)$.

Now we shall assume that G is reductive and X is complete. As before, B will denote a Borel subgroup in G containing a maximal torus T and $\lambda : k^* \to T$ be a regular one-parameter subgroup in T having the same fixed points on X as T. Let us assume that the action of $\lambda(k^*)$ on X is locally linear (cf [4]). Denote by X^+ the sink of this action. This is a connected and irreducible component in $X^{\lambda(k^*)}$ ([4], Theorem 9). From Theorem 2 it follows

16. Proposition. GX^+ is a connected and irreducible component in X^*.

Proof. X^+ is a minus cell of X ([4], Remark 11). By Remark 3 $X^+ \subset X^{B^-(\lambda)}$. Since $X^{B^-(\lambda)} \subset X^{\lambda(k^*)}$ and X^+ is a component in $X^{\lambda(k^*)}$, then X^+ is a component in $X^{B^-(\lambda)}$. Therefore GX^+ is a connected and irreducible component in

$X^* = GX^{B^-(\lambda)}$.

Let $X_1^*, \ldots, X_r^*$ denote all connected components of X^*.

17. Definition. For $i = 1, \ldots, r$ let $X_i = \{x \in X : \text{sink } \overline{Gx} \subset X_i^*\}$.

Then $X_i \supset X_i^*$ for all i, $X_i \cap X_j = \emptyset$ for $i \neq j$ and X is the union of the X_i's. Obtained decomposition will be called the decomposition of X determined by the action of G. Notice, that it depends on a choice of λ. We call X_i the cells of this decomposition. So now we have two types of cells - plus and minus cells determined by the action of $\lambda(k^*)$ and the cells defined above.

Let X_d be the cell containing $GX^+ = X_d^*$. Then $X^+ = (X_d^*)^{B^-(\lambda)}$. For $x \in X_d$, $\text{sink } Gx \subset X^{B^-(\lambda)}$, hence $X_d = \{x \in X : \text{sink } \overline{Gx} \subset X^+\}$. Denote the big plus cell of the action of $\lambda(k^*)$ on X by U. Then $X_d = \{x \in X : Gx \cap U \neq \emptyset\} = GU = \bigcup_{g \in G}$ (big cell of the action of $g\lambda(k^*)g^{-1}$ on X). The last equality follows, since $\lim_{t \to 0} \lambda(t)x = y$ if and only if $\lim_{t \to 0} g\lambda(t)g^{-1}gx = gy$ for $g \in G$, $x, y \in X$. X_d is called the big cell.

18. Corollary. The big cell X_d is open in X.

Next, let us consider other cells. Choose one of them, e.g. X_1. For $y \in X_1$, $\text{sink } \overline{Gy} \subset X_1^{B^-(\lambda)}$, hence the big cell of the action of $\lambda(k^*)$ on $\overline{Gy}$ is contained in $\overline{X^+(X_1^{B^-(\lambda)})}$. So, $Gy \subset \overline{Gy} \subset \overline{X^+(X_1^{B^-(\lambda)})}$. This inclusion holds for all $y \in X_1$, whence $X_1 \subset \overline{X^+(X_1^{B^-(\lambda)})}$.

19. Theorem. The cells defined above are locally closed in X, if the action of $\lambda(k^*)$ has the property A) : the plus cells are locally closed and the projections along orbits are regular.

Proof. Suppose X_1 is not locally closed in a point $z \in X_1$. Since $\{y \in X_1 : X_1$ is not locally closed at $y\}$ is closed in X_1 and G-invariant, we may assume $z \in X_1^{B^-(\lambda)}$. Then in some neighbourhood of z in $\overline{X_1}$ there exists a point lying not in X_1. We may take $X^+(X_1^{B^-(\lambda)}) \cap \overline{X_1}$ as such neighbourhood, it is open by A). Let y be an element of $\overline{X_1} \cap X^+(X_1^{B^-(\lambda)}) - X_1$. Then $Gy \cap X^+(X_1^{B^-(\lambda)})$ is open in Gy, since $Gy \subset \overline{X_1} \subset \overline{X^+(X_1^{B^-(\lambda)})}$ and $Gy \cap X^+(X_1^{B^-(\lambda)}) \neq \emptyset$. Therefore $\text{sink } Gy \subset X_1^{B^-(\lambda)}$, thus $y \in X_1$, a contradiction. Hence the cell X_1 is locally closed.

20. Corollary. In particular, if X is nonsingular or normal projective, then all the cells are locally closed.

Now we shall describe the cells containing fixed points.

21. Example. If G is semisimple and a component X_2^* of X^* consists of fixed points, then $X_2 = X_2^*$.

Proof. We may assume X normal. Let x be a point in the cell X_2 and let Y denote the sink of the action of $\lambda(k^*)$ on $\overline{Gx}$. Then, for $g \in G$, the sink of the action of $g\lambda(k^*)g^{-1}$ on $\overline{Gx}$ is the set gY. Since $Y \subset X^G$, Y is the sink of the action of any torus conjugate to $\lambda(k^*)$ on $\overline{Gx}$. For $w \in W$, the Weyl group, let $w(\lambda)$ be a one-parameter subgroup in T given by $w(\lambda)(t) = n\lambda(t)n^{-1}$ for $t \in k^*$, where $n \in N(T)$ represents $w \in W$. For $w \in W$, denote by U_w the big cell of the action of $w(\lambda)$ on $\overline{Gx}$ (the sink is Y) and by $\pi_w : U_w \to Y$ the projection along orbits of this action. Choose a point $y \in Y$. Since X is normal, there exists a T-invariant open affine neighbourhood V of y in $\overline{Gx}$. Since Gx and each of the sets $\pi_w^{-1}(V \cap Y), w \in W$, are open in $\overline{Gx}$, then the set $U = Gx \cap \bigcap_{w\in W} \pi_w^{-1}(V \cap Y)$ is nonempty. Let $z \in U$. Then, for $w \in W$, we have $\lim_{t\to 0} w(\lambda)(t)z \in V \cap Y$. Embed V T-equivariantly in an affine space A with a linear action of T. Let Z be the component of A^T such that $V \cap Y \subset Z$. In particular, for $w \in W$, $\lim_{t\to 0} w(\lambda)(t)z \in Z$. Let $\mu = \sum_{w\in W} w(\lambda)$, i.e. $\mu(t) = \prod_{w\in W} w(\lambda)(t)$ for $t \in k^*$. Since the weights of the action of $\mu(k^*)$ are sums of the corresponding weights of the actions of $w(\lambda)$, $\lim_{t\to 0} \mu(t)z \in Z$. We shall show that μ is trivial, i.e. $\mu(k^*) = \{e\}$. Then it follows $z \in Z$, so it is fixed by T. z was arbitrary, whence the action of T is trivial on U. Thus it is trivial on $\overline{U} = \overline{Gx}$. Since Y is a component of $\overline{Gx}^T$, then $Y = \overline{Gx}$. In particular $x \in X^G$. It remains to prove $\mu(k^*) = \{e\}$. Notice that μ is a fixed point of the action of W on $Y(T)$. Indeed, for $w' \in W$, $w'(\mu)(t) = w' \prod_{w\in W} w(\lambda)(t) = \prod_{w\in W} w'w(\lambda)(t) = \mu(t)$. Thus, $\mu(k^*)$ is contained in the connected component $(T^W)^0$ of the set of fixed points of the action of W on T. Since G is semisimple, then $(T^W)^0 = \mathrm{Rad}\, G = \{e\}$ by [2], 14.2. Therefore $\mu(k^*) = \{e\}$ and we are done.

22. Corollary. [1] If a semisimple group acts nontrivially on a complete variety X, then there exist in X nontrivial projective orbits.

23. Example. (cf [1]) Assume G is semisimple and there is given a representation $\rho = \rho_1^{m_1} \oplus \ldots \oplus \rho_k^{m_k}$ of G in a vector space $V = \oplus V_i^{m_i}$, where ρ_i are irreducible representations in V_i for $i = 1,\ldots,k$. This induces an action of G on $\mathbb{P}(V)$. Then, as shows an easy computation, the cells of the decomposition of $\mathbb{P}(V)$ determined by this action are of the form $\mathbb{P}(V)_i = \mathbb{P}(V_1^{m_1} \oplus \ldots \oplus V_i^{m_i}) - \mathbb{P}(V_1^{m_1} \oplus \ldots \oplus V_{i-1}^{m_{i-1}})$ for $i = 2,\ldots,k$ and $\mathbb{P}(V)_1 = \mathbb{P}(V_1^{m_1})$.

REFERENCES

[1] A. Bialynicki-Birula, On action of SL(2) on complete algebraic varieties, Pac. J. of Math. 86 (1980), 53-58.

[2] A. Borel, Linear algebraic groups, W.A. Benjamin 1969.

[3] J.B. Carrell, A.J. Sommese, SL(2) actions on compact Kaehler manifolds, I.

[4] J. Konarski, A pathological example of an action of k^*, this volume.

[5] H. Sumihiro, Equivariant completion, J. Math. Kyoto Univ. 14 (1974).

Institute of Mathematics, Warsaw University
PKiN, 00-901 Warszawa

LINEARIZATION OF REDUCTIVE GROUP ACTIONS

by

Mariusz Koras

The purpose of this paper is to prove the following theorem

Theorem 1. Let G be a connected, reductive, complex Lie group. Let $\sigma : G \times X \to X$ be an analytic action on a compact Kähler manifold X. Then for every point $p \in X^G$ there exist a G-invariant open neighborhood U of p and an analytic isomorphism $\phi : U \to \phi(U) \subset C^n$ such that $\phi(p) = 0$ and ϕ is G-equivariant with respect to a linear action of G on C^n.

In connection with the above see [3], [4] where a similar theorem is proved in the infinitesimal version.

Definition 1. Let $\sigma : G \times X \to X$ be a continuous action of a topological group G on a topological space X. We say that σ has the C-property at a point $p \in X^G$ iff the following condition holds: for every open neighborhood U of p there exists an open neighborhood $V \subset U$ of p such that for every point $x \in X$ the set $\{g \in G : gx \in V\}$ is connected.

We shall show that this property is equivalent to the linearization of an analytic action of a connected reductive group on a complex manifold near a fixed point.

Let G be as in Theorem 1. Let Z denote the centre of G and Z_o the connected component of the unit in Z. Z_o is a torus, i.e. is isomorphic to a product $C^* \times C^* \times \ldots \times C^*$, and $G = Z_o \cdot P$ where P is a semisimple, connected group. Let K be a maximal, compact, connected subgroup in P. Lie algebras of G, P, $K \ldots$ we shall denote by $L(G)$, $L(P)$, $L(K) \ldots$. Then $L(P) = L(K) + iL(K)$. Let T be a maximal torus in G and K_1 be the maximal, connected, compact, commutative subgroup in K contained in T. Let $iv \in iL(K)$. There exists a maximal, connected, compact, commutative subgroup of the group K containing $\exp v$. This subgroup is conjugate in K with K_1. It follows that there exists a $k_o \in K$ such that $k_o . \exp v . k_o^{-1} \in T$. Thus $v \in L(k_o^{-1}.T.k_o)$. Also $iv \in L(k_o^{-1}.T.k_o)$ This implies that $\exp iv \in k_o^{-1}.T.k_o$. It is known that $P = K.\exp(iL(K))$. Therefore, if $g \in G$ then $g = z_o.p$, $z_o \in Z_o$, $p \in P$ and $p = k.\exp iw$ for some $k \in K$ and $w \in L(K)$. There exist $k_o \in K$, $t \in T$ such that $\exp iw = k_o^{-1}.t.k_o$. Thus $g = z_o.k.k_o^{-1}.t.k_o = k.k_o^{-1}.z_o.t.k_o$. But Z_o is contained in every maximal torus of G. Thus we have proved the following lemma

Lemma 1. $G = K . \bigcup_{k_o \in K} k_o.T.k_o^{-1}$

Lemma 2. Let $T = \underbrace{C^* \times \ldots \times C^*}_{m}$ be a torus. Let $\sigma : T \times C^n \to C^n$ be a linear action (analytic) of T. Then σ has the C-property at $0 \in C^n$.

Proof. Let $S = \underbrace{S^1 \times \ldots \times S^1}_{m}$ be the maximal, connected, compact subgroup of T. We may assume that T acts diagonally; this means that there exist integers k_{ij} such that $(t_1,\ldots,t_m)\cdot(z_1,\ldots,z_n) = (t_1^{k_{11}}\cdot\ldots\cdot t_m^{k_{m1}} z_1,\ldots, t_1^{k_{1n}}\cdot\ldots\cdot t_m^{k_{mn}}\cdot z_n)$. Let $U_\varepsilon = \{(z_1,\ldots,z_n) : |z_i| < \varepsilon,\ i = 1,\ldots,n\}$. Since U_ε is S-invariant it is enough to show that the set $\{t \in R^+ \times \ldots \times R^+ : tx \in U_\varepsilon\}$ is connected for every $x \in C^n$. But this set is equal to the set $\bigcap_{i=1}^{n}\{(t_1,\ldots,t_m) \in R^+ \times \ldots \times R^+ : t_1^{k_{1i}}\cdot\ldots\cdot t_m^{k_{mi}} < \frac{\varepsilon}{|x_i|}\}$ which is connected.

Proposition 1. Let T be a torus and $\sigma : T \times X \to X$ be a holomorphic action on a complex manifold. Assume that σ has the C-property at a point $p \in X^T$. Then σ is linearizable near p i.e. there exists a T-invariant open neighborhood U of p and an analytic isomorphism $\phi : U \to \phi(U) \subset C^n$ such that $\phi(p) = 0$ and ϕ is T-equivariant with respect to a linear action of T on C^n.

Remark 1. Let G be a complex, reductive, connected Lie group and K denote a maximal, connected, compact subgroup of G. Let $\sigma : G \times X \to X$ be an analytic action of G on a complex manifold X. Let $p \in X^G$ be a fixed point of this action. It is known (see [3], [4]) that there exists an arbitrarily small open K-invariant neighborhood U of p and an analytic coordinate system on U such that K acts linearly on U (with respect to these coordinates).

Proof of Proposition 1. Assume $T = C^* \times \ldots \times C^*$ and $S = S^1 \times \ldots \times S^1 \subset T$. It follows from the above remark that there exists an open S-invariant neighborhood V of p and an analytic isomorphism $\phi : V \to \phi(V) \subset C^n$ such that $\phi(p) = 0$ and ϕ is S-equivariant with respect to a linear action of S on C^n. We may extend the action of S on C^n to the analytic linear action of T. We may assume that each orbit of T intersects the set $\phi(V)$ along a connected set (lemma 2). Let $\psi = \phi^{-1}$. We extend ψ to $\tilde{\psi} : \bigcup_{t\in T} t\phi(V) \to X$ letting $\tilde{\psi}(tx) = t\psi(x)$ for $t \in T$ and $x \in \phi(V)$. We must check that this is well defined. Let $A = \{t\in T : tx\in\phi(V)\}$ for fixed $x \in \phi(V)$. The set A is connected and contains S. Let us consider two analytic maps $\alpha,\beta : A \to X$ where $\alpha(t) = t\psi(x)$ and $\beta(t) = \psi(tx)$. They are equal on S, thus they are equal. It follows easily that $\tilde{\psi}$ is well defined. Now choose $V_1 \subset V$ an open connected neighborhood of p such that the set $\{t\in T : tx \in V_1\}$ is connected for every $x \in X$. As above we can extend $\phi : V_1 \to C^n$ to $\tilde{\phi} : \bigcup_{t\in T} tV_1 \to C^n$. $\tilde{\psi}$ considered on $\bigcup_{t\in T} t\phi(V_1)$ is the inverse map to $\tilde{\phi}$ because $\tilde{\phi} \circ \tilde{\psi} = \mathrm{id}$ on $\phi(V_1)$. Of course $\tilde{\phi}$ is T-invariant.

Proposition 2. Let G be as in Theorem 1. Let $\sigma : G \times X \to X$ be an analytic action of G on a complex manifold X. Let T be a maximal torus of G. Assume that the induced action of T on X has the C-property at a point $p \in X^G$. Then σ has the C-property at p.

Proof. Let K be as in Lemma 1. For every $k \in K$, let $\sigma_k : T \to \text{Aut } X$ be defined by $\sigma_k(t)x = \sigma(ktk^{-1},x)$. We shall prove that for every open neighborhood U of p there exists a K-invariant open neighborhood V of p such that $V \subset U$ and for every $k \in K$, $x \in X$ the set $\{t\in T : \sigma_k(t)x \in V\}$ is connected. We may assume that U is K-invariant and there exists an analytic isomorphism $\phi : U \to \mathbb{C}^n$, $\phi(p) = 0$ such that ϕ is K-invariant with respect to some linear action at K on C^n (remark 1). We may assume that K acts on C^n by isometries and T acts diagonally. We may also assume that for every ε the set $B_\varepsilon = \{(z_1,\dots,z_n) : |z_i| < \varepsilon,\ i=1,\dots,n\}$ is contained in $\phi(U)$. Put $U_n = \phi^{-1}(B_{\frac{1}{n}})$; U_n is K-invariant. Let us assume that for every n there exist $k_n \in K$ and $x_n \in U_n$ such that the set $\{t\in T : \sigma_{k_n}(t)\, x_n \in U_n\}$ is disconnected. The action of the torus T has the C-property at p, so there exists an open neighborhood $W \subset U$ of p that the set $\{t\in T : tx \in W\}$ is connected for every $x \in X$. As in the proof of Proposition 1 the isomorphism $\phi : W \to \phi(W)$ extends to T-invariant isomorphism $\tilde{\phi} : \bigcup_{t\in T} tW \to \bigcup_{t\in T} t\phi(W)$. From the proof of Lemma 2 it follows that, for every $U_k \subset W$ and $x \in X$, the set $\{t\in T : tx \in U_k\}$ is connected. Let $U_r \subset W$. Put $y_r = k_r^{-1}x$. Then the set $\{t\in T : \sigma_{k_r}(t)x_r \in U_r\} = \{t : (k_r t k_r^{-1})x_r \in U_r\} = \{t : tk_r^{-1}x_r \in U_r\} = \{t : ty_r \in U_r\}$ is connected, a contradiction.

Now let U be an arbitrary open neighborhood of p. Let $V \subset U$ be a K-invariant open neighborhood of p such that for all $k \in K$ and all $x \in X$ the set $\{t\in T : \sigma_k(t)x \in V\}$ is connected. Assume that $x \in V$ and $gx \in V$ for some $g \in G$. Decompose g as $g = k.k_o.t.k_o^{-1}$, $k,k_o \in K$, $t \in T$ (lemma 1). Since $kV = V$, $(k_o.t.k_o^{-1})x = \sigma_{k_o}(t)x \in V$. There exists a path $\gamma : \langle 0,1\rangle \to T$ such that $\gamma(0) = 1, \gamma(1) = t$ and for every $r \in \langle 0,1\rangle$ $(k_o.\gamma(r).k_o^{-1})x \in V$. Let $\alpha : \langle 0,1\rangle \to K$ be a path joining 1 with k. Put $\beta(r) = \alpha(r).k_o.\gamma(r).k_o^{-1}$. Then $\beta(0) = 1$, $\beta(1) = g$ and for every $r \in \langle 0,1\rangle$ $\beta(r)x \in V$. It follows that the set $\{g\in G : gx \in V\}$ is connected for every $x \in X$.

It follows from Proposition 2 and Lemma 2 that every linear action of a connected, reductive group on affine space C^n has the C-property at 0. Repeating the proof of Proposition 1 we get the following theorem:

Theorem 2. Let G be as in Theorem 1. Let $\sigma : G \times X \to X$ be an analytic action of G on a complex manifold X. Then σ is linearizable in a neighborhood of a point $p \in X^G$ iff it has the C-property at p.

For the proof of Theorem 1 we need the following theorem due to Frankel.

Theorem 3 (Frankel [2]). Let $\{\phi_t\}t \in R$ be a 1-parameter group of isometries acting on a compact manifold X. Let V be the vector field induced by this group. Assume that the zero set of V is nonempty. If ω denotes the Kähler form on X, then there exists a smooth real function f on X such that $i(V)\omega = df$, where $i(V)$ denotes the contraction operator induced by V.

Assume we have an analytic action of C^* on a compact Kähler manifold X. Assume that the Kähler form on X is S^1-invariant and $X^{C^*} \neq \emptyset$. Then, from Frankel's theorem, there exists a corresponding function. Every such function we call a Frankel function of this action. It is a very nice (and easy) fact observed by Carrell and Sommese [1] that a Frankel function is strictly increasing along the nontrivial orbits of $R^+ \subset C^*$.

Proof of Theorem 1. It is enough to show that an action of a torus T on a compact Kähler manifold X has the C-property at each fixed point. Let $p \in X^T$. Assume $T = C^* \times \ldots \times C^*$ and let $S = S^1 \times \ldots \times S^1$. Then $L(T) = L(S) + i\,L(S)$. We may assume that the Kähler form on X is S-invariant. There exists the 1-1 correspondence between elements of $iL(S)$ and 1-parameter subgroups $\gamma : R^+ \to T$ defined by $\gamma(t) = (t^{a_1},\ldots,t^{a_m}), a_i \in R$. In the following we shall identify an element of $iL(S)$ corresponding to γ with the sequence $A = (a_1,\ldots,a_m)$. This identification gives an R-isomorphism $iL(S) \overset{\to}{\sim} R^m$. For $A \in iL(S)$ we denote by σ_A the induced action of R^+ on X. Let f_A be the corresponding Frankel function. f_A is strictly increasing along the nontrivial orbits of the action σ_A. We may choose functions f_A, $A \in iL(S)$ in such a way that the function $f : iL(S) \times X \to R$ defined by $f(A,x) = f_A(x)$ is smooth. Let U be an S-invariant open neighborhood of p and $\phi : U \to \phi(U) \subset C^n$ be an analytic isomorphism such that $\phi(p) = 0$ and ϕ is S-equivariant with respect to a linear action of S on C^n. We may assume that S acts on C^n by isometries and we may extend this action to the linear action of T on C^n. We may also assume that the action is diagonal. Thus there exist integers k_{ij} such that $(t_1,\ldots,t_m)(z_1,\ldots,z_n) = \left(\ldots, t_1^{k_{1k}} \cdot \ldots \cdot t_m^{k_{mi}} \cdot z_i,\ldots\right)$ for every $(t_1,\ldots,t_m) \in T$ and $(z_1,\ldots,z_n) \in C^n$. Let $A_k = \{z \in C^n : |z_i| < \frac{1}{k},\ i = 1,\ldots,n\}$. Let $\|\cdot\| : iL(S) \to R$ denote the euclidean norm on R^m. We shall prove that for all n there exists a $k \geq n$ such that for all $x \in X$ and for all $A \in iL(S)$ with $\|A\| = 1$, the set $\{t \in R^+ : \sigma_A(t)x \in \phi^{-1}(A_k)\}$ is connected. Assume not. Then there exists an n such that for all $k \geq n$, there exists an $x \in X$ and $A \in iL(S)$ with $\|A\| = 1$ so that the set $\{t \in R^+ : \sigma_A(t)x \in \phi^{-1}(A_k)\}$ is disconnected. Thus we obtain sequences $(x_k)_{k\geq n}$, $(A_k)_{k\geq n}$, $x_k \in X$, $A_k \in iL(S)$ with $\|A_k\| = 1$ such that the set $D_{k,x_k} = \{t \in R^+ : \sigma_{A_k}(t)x_k \in \phi^{-1}(A_k)\}$ is disconnected for every $k \geq n$. We may assume that $A_k \to A_0$. Let a_k, b_k be parameters such that $\sigma_{A_k}(a_k)x_k \in \phi^{-1}(A_k)$, $\sigma_{A_k}(b_k)x_k \in \phi^{-1}(A_k)$ and the interval $[a_k,b_k]$ is not contained in D_{k,x_k}. Assume that $a_k > b_k$. Set $\sigma_{A_k}(a_k)x_k = y_k$, and $b_k/a_k = c_k$. Then $y_k \in \phi^{-1}(A_k)$, $\sigma_{A_k}(c_k)y_k \in \phi^{-1}(A_k)$ and the interval $[c_k,1]$ is not contained in $D_{k,y_k} = \{t \in R^+ : \sigma_{A_k}(t)y_k \in \phi^{-1}(A_k)\}$. The isomorphism ϕ has the property that if z, $t_0 z \in U$ and $1, t_0$ belong to the same connected component of the set $\{t \in T : tz \in U\}$ then $\phi(t_0 z) = t_0\phi(z)$. As D_{k,y_k} is disconnected,

$\{t\in R^+ : \sigma_{A_k}(t)y_k \in \phi^{-1}(A_1)\}$ is disconnected (this follows from the arguments used in the proof of Lemma 2). Hence, for every $k \geq n$ there exists a parameter $d_k \in R^+$ such that $c_k < d_k < 1$ and $\sigma_{A_k}(d_k)y_k \notin \phi^{-1}(A_1)$. Assume that the sequence (d_k) has a limit point $d \neq 0$. Let $d_{k_1} \to d$. Then $\sigma_{A_{k_1}}(d_{k_1})y_{k_1} \to \sigma_{A_o}(d)p = p$. But $\sigma_{A_{k_1}}(d_{k_1})y_{k_1} \in X - \phi^{-1}(A_1)$ for every k_1. Therefore p would be in $X - \phi^{-1}(A_1)$. Thus $d_k \to 0$ and also $c_k \to 0$. Let $q \in X - \phi^{-1}(A_1)$ be a limit point of the sequence $\sigma_{A_k}(d_k)y_k$.

In the product $R \times R^m \times X \times X \times R \times X$ we consider the set $\{(t,A,x,y,s,z) : y = \sigma_A(t)x,\ z = \sigma_A(s)x,\ 0 < t < s < 1\}$, a semianalytic set. The point $(0,A_o,p,p,0,q)$ belongs to the closure of this set. Thus we can reach this point from that set with the aid of an R-analytic curve. More precisely, there exist R-analytic curves defined on $[0,\varepsilon]$:

(1)	$t(r)$	$t(0) = 0$	$t(r) \in R$
(2)	$A(r)$	$A(0) = A_o$	$A(r) \in R^m = iL(S)$
(3)	$x(r)$	$x(0) = p$	$x(r) \in X$
(4)	$y(r)$	$y(0) = p$	$y(r) \in X$
(5)	$s(r)$	$s(0) = 0$	$s(r) \in R$
(6)	$z(r)$	$z(0) = q$	$z(r) \in X$

such that

(7) $0 < t(r) < s(r) < 1$ for $r > 0$

(8) $y(r) = \sigma_{A(r)}(t(r))x(r)$, $z(r) = \sigma_{A(r)}(s(r))x(r)$ for $r > 0$

It follows from (8), (4), (6) that the set $\{t\in R^+ : \sigma_{A(r)}(t)x(r) \in \phi^{-1}(A)$ is disconnected for $r < \varepsilon' < \varepsilon$. Let $p(r)$ be the right end point of connected component of parameter $t(r)$ in the set $\{t\in R^+ : \sigma_{A(r)}(t)x(r) \in \phi^{-1}(A_2)\}$, $r < \varepsilon'$. Of course $t(r) < p(r) < 1$. We introduce analytic coordinates $z = (z_1,\dots,z_n)$ on $\phi^{-1}(A_1)$ with the aid of ϕ. Let $A(r) = (a_1(r),\dots a_m(r))$, $A_o = (a_1,\dots,a_m)$. It follows from the definition of $p(r)$ that $\sigma_{A(r)}(p(r))\, x(r) \in \overline{\phi^{-1}(A_2)} - \phi^{-1}(A_2)$. We may assume that $\sigma_{A(r)}(p(r))\, x(r) \to w = (w_1,\dots,w_n)$. Notice that

$$\sigma_{A(r)}(p(r))\, x(r) = \sigma_{A(r)}\left(p(r)/t(r)\right) y(r) =$$

$$\left(\dots, \left(p(r)/t(r)\right)^{k_{1i}a_1(r)+\dots+k_{mi}a_m(r)} \cdot y_i(r), \dots\right)$$

We have (9) $1 < p(r)/t(r) < 1/t(r)$ and $y_i(r) \to 0$, $i = 1,\dots,n$. If $\sum_{j=1}^m k_{ji}a_j < 0$, then

$$\left(u_{i(r)} = \left(p(r)/t(r)\right)^{\sum_{j=1}^m k_{ji}a_j(r)} \cdot y_i(r) \to 0 .\right.$$

Thus, in this case $w_i = 0$. Assume that $\sum_{j=1}^m k_{ji_o}a_j = 0$. If there exists a sequence $r_n \to 0$ such that $\sum_j k_{ji_o}a_j(r_n) \leq 0$ for every n, then $u_{i_o}(r_n) \to 0$

and $w_{i_o} = 0$ again. Assume now that $\sum_j k_{ji_o} a_j(r) > 0$ for every r. It follows from (4), (1) that $y_{i_o}(r) = r^{\ell}\cdot\tilde{y}_{i_o}(r)$, $\ell > 0$, $\tilde{y}_{i_o}(0) \neq 0$ and $t(r) = r^u.\tilde{t}(r)$, $\tilde{t}(0) \neq 0$, $u > 0$. Hence, from (9), we obtain

$$u_{i_o}(r) < t(r)^{-\sum_j k_{ji_o} a_j(r)} . |y_{i_o}(r)| =$$

$$r^{-u \sum_j k_{ji_o} a_j(r) + 1} . \tilde{t}(r)^{-\sum_j k_{ji_o} a_j(r)} . \left|y_{i_o}(r)\right| \to 0 \quad \text{as} \quad r \to 0$$

We have again obtained that $w_{i_o} = 0$. Thus we have proved that $w_i = 0$ provided $\sum_{j=1}^{m} k_{ji} a_j \leq 0$.

Since the action σ_{A_o} satisfies $\sigma_{A_o}(t)z = \left(\ldots, t^{\sum_j k_{ji} a_j} .z_i, \ldots\right)$ for every $z \in \phi^{-1}(A_1)$ it follows that $\lim_{t \to 0} \sigma_{A_o}(t)w = p$. Moreover $w \neq p$ because $w \in \phi^{-1}(A_2) - \phi^{-1}(A_2)$. It follows that w is not a fixed point of the action σ_{A_o}. Since $t(r) < p(r) < 1$,

$$f_{A(r)}\left(\sigma_{A(r)}(t(r))\, x(r)\right) < f_{A(r)}\left(\sigma_{A(r)}(p(r))\, x(r)\right) <$$

$$< f_{A(r)}(x(r)) .$$

Taking the limit as $r \to 0$, we obtain $f_{A_o}(p) \leq f_{A_o}(w) \leq f_{A_o}(p)$. Thus $f_{A_o}(w) = f_{A_o}(p)$. But this is impossible because f_{A_o} is strictly increasing along the orbit $\sigma_{A_o}(t)w$. It follows that for all n, there exists a $k \geq n$ such that for all $x \in X$ and $A \in iL(S)$, the set $\{t \in R^+ : \sigma_A(t)\, x \in \phi^{-1}(A_k)\}$ is connected.

Assume now that the set $\{t \in R^+ : \sigma_A(t)\, x \in \phi^{-1}(A_k)\}$ is connected for every $x \in X$ and $A \in iL(S)$. Let x, $t_o x \in \phi^{-1}(A_k)$, $t_o \in T$. We can find $s \in S$ and $A \in iL(S)$ such that $t_o = s.\sigma_A(t)$. The set $\phi^{-1}(A_k)$ is S-invariant hence $\sigma_A(t)\, x \in \phi^{-1}(A_k)$. There exists a path $\alpha : [0,1] \to R^+$ such that $\alpha(0) = 1$, $\alpha(1) = t$ and $\sigma_A(\alpha(r))\, x \in \phi^{-1}(A_k)$ for $r \in [0,1]$. Let $\beta : [0,1] \to S$ be a path joining 1 with s. Take $\tau(r) = \beta(r)\sigma_A(\alpha(r))$. Then $\tau(0) = 1$, $\tau(1) = t_o$ and $\tau(r)x \in \phi^{-1}(A_k)$ for every $r \in [0,1]$.

We have proved that for all n, there exists a $k \geq n$ so that for all $x \in X$ the set $\{t \in T : tx \in \phi^{-1}(A_k)\}$ is connected. Q.E.D.

REFERENCES

1. J. Carrell, A. Sommese : Some topological aspects of C*-actions on compact Kaehler manifolds, Comment. Math. Helvetici 54 (1979).

2. T. Frankel, Fixed points on Kaehler manifolds, Annals of Math. 70(1959).

3. V. Guillemin, S. Sternberg, Remarks on a paper of Hermann, Trans. Amer. Math. Soc. 130(1968).

4. A.G. Kusznirenko, An analytic action of a semisimple group is equivalent to a linear action near fixed point (Russian) Functional Analysis 1 (1967)

HOLOMORPHIC VECTOR FIELDS AND RATIONALITY

by

David I. Lieberman*

Preface.

This manuscript was written in the fall of 1973. It was not published, since the referee pointed out that the main result (rationality of a projective manifold having a holomorphic vector field with a generic zero) could be understood more directly by employing techniques from the theory of algebraic groups. I have carried out this program subsequently (cf. Proc. Sympos. Pure Math IXXX, p. 273). However, the techniques, notably equivariant projection and the theory of substantial vector fields, developed in the present manuscript have an independent interest and provide a ready reference for the study of equivariant geometry. In view of the number of requests I have received for the manuscript and the number of occasions it has been cited as an (unpublished) reference, I am pleased to have this opportunity to include it in these proceedings. Perhaps it may prove of use in settling the still unsolved conjecture that a projective manifold having a holomorphic vector field with isolated zeroes is rational.

Introduction.

This note is a preliminary study of Carrell's problem: given a compact complex space X having a global holomorphic vector field with isolated zeroes determine whether X must be rational (bimeromorphic to $\mathbb{P}^n$). All spaces studied are assumed reduced and irreducible.

A compact Kahler n-manifold X having a global holomorphic vector field V with isolated zeroes is known to be projective algebraic and to have vanishing irregularity (i.e. $H^0(X, \Omega^1_X) = 0$,) [5]. Moreover, the plurigenera $P_r = \dim(H^0(X, (\Omega^n_X)^{\otimes r}))$, $r > 0$, are known to vanish [10]. Thus if $n = 0$, 1 or 2 then X is known to be rational (cf. also [4]).

One can easily construct examples of non-rational normal surfaces having vector fields with isolated zeroes (cf. §2 below). When one resolves the singular points on such surfaces the vector field lifts uniquely to the resolution but will have non-isolated zeroes. To avoid this disappearance of isolated zeroes of a vector field under blowing up we introduce the notion of "generic zero". Namely, given any (possibly singular) algebraic variety X and a global holomorphic vector field V vanishing at $x \in X$, then to say x is generic means roughly that given any $\hat{X} \xrightarrow{\pi} X$ obtained by a succession of monoidal transforms such that V lifts to $\hat{V}$

*(Sloan Foundation Fellow, partially supported by National Science Foundation Grant GP-28323A2.)

on $\hat{X}$ then $\hat{V}$ has isolated zeroes on $\pi^{-1}(x)$. (In fact these zeroes of $\hat{V}$ are again evidently generic). Our main theorem is

Theorem 1: If X is a projective algebraic variety over $\mathbb{C}$ and V is a global holomorphic vector field on X having a generic zero, then X is rational.

As a corollary one obtains

Theorem 2: If X is a compact Kahler manifold and V is a global holomorphic vector field having a generic zero and satisfying $\dim(\text{zero}(V)) \leq 1$, then X is rational.

This theorem follows from the remark in [5] that such Kahler manifolds are all algebraic. The theorems can be immediately generalized to Moisezon spaces by proving an equivariant version of the Chow lemma, (a result claimed by Hironaka).

The study begins by collecting, in §1, the basic results about global vector fields and equivariant maps. Given a vector field V on X, a map $f : X \longrightarrow Y$ is called V-equivariant if there exists a vector field W on Y so that $W = f_*(V)$. By abuse of language, if only V (resp. W) is given, f will be called equivariant provided some choice of W (resp. V) renders f equivariant. Key results of §1 concern substantial vector fields. Substantiality means the only global meromorphic functions annihilated are constants. Every vector field having a generic zero is substantial. The property of substantiality is preserved under equivariant birational maps, and more importantly if V is substantial on X and $f : X \longrightarrow Y$ is a V-equivariant rational map with dense image, then $f_*(V)$ is substantial on Y.

In §2 we show that all the standard rational varieties: flag manifolds and Grassmanians and products thereof, all carry substantial vector fields and we classify the substantial vector fields on $\mathbb{P}^n$. We also exhibit examples of nonrational normal surfaces having global (insubstantial) vector fields with isolated zeroes. We know of no example of a nonsingular nonrational projective variety having a global vector field with only isolated zeroes.

In §3 we show that every projective variety X of dimension n carrying a holomorphic substantial vector field with $\text{zero}(V) \neq \phi$ admits an equivariant rational map $g : X \longrightarrow \mathbb{P}^n$ with dense image. The argument is by induction on n and proceeds by equivariantly imbedding $X \longrightarrow \mathbb{P}^N$ (Blanchard's theorem) and employs the technique of equivariant projection. This technique was inspired by work of A. Howard [9].

There exists an invariant hypersurface $H \subseteq \mathbb{P}^n$ such that the map $X - g^{-1}(H) \longrightarrow \mathbb{P}^n - H$ is an unramified finite covering (cf. §4). The key argument of the paper is to show that the only invariant hypersurfaces in $\mathbb{P}^n$ are unions of coordinate hyperplanes, which follows from the substantiality of the vector field on $\mathbb{P}^n$, (recall the argument of §1). But then $X - g^{-1}(H)$ is a cover of either (a) $\mathbb{P}^n$ (if $H = \phi$), (b) $\mathbb{C}^n$, or (c) $\mathbb{C}^* \times \ldots \times \mathbb{C}^* \times \mathbb{C}^k$. In cases (a) and (b) the covering is clearly one sheeted and $\hat{X}$ is rational. In case (c) the covering may have many sheets, but the theory of the algebraic π_1 (cf. [7]) implies $\hat{X} - g^{-1}(H) \xrightarrow{\sim} \mathbb{C}^* \times \ldots \times \mathbb{C}^* \times \mathbb{C}^k$ (by an algebraic map) and again X is rational.

The hypotheses of the theorem should clearly be weakened. The characteristic 0 hypothesis seems essential only for the application of Hironaka's work. The substantiality hypothesis (i.e. generic zeroes) is only essential in §4 where it is employed to restrict the nature of the "branch locus" H. Various other hypotheses (e.g. $H^0(X,\Theta) \neq 0$ and $H^0(X,\Omega^1) = 0$) would suffice to find the equivariant rational $g : X \longrightarrow \mathbb{P}^n$ by induction. A more careful analysis of hypotheses needed to restrict the nature of H is clearly required.

The problem of studying the relationship of existence of isolated zeroes of vector fields and rationality questions was proposed by J. Carrell to the author. After several unsuccessful joint attempts on the problem by rather different techniques, we set it aside. The author owes both his interest in the problem and his knowledge of basic techniques to the stimulating collaboration with Carrell at Purdue. Theorems 1.1 - 1.6 were obtained jointly. The present work would not have been possible without Hironaka's gracious explanations of his recent, as yet unpublished works. Thanks are also due W. Messing for a willing ear and for his commuter course on π_1.

§1. Equivariant maps.

Given a complex space X and $V \in H^0(X,\Theta)$, a global holomorphic vector field, then general holomorphic maps $f : X \longrightarrow Y$ or $g : Y \longrightarrow X$ will not be equivariant, i.e. V will not push forward with f or lift with g. For example, a subvariety Z of X is equivariantly embedded only if V is tangent to Z. Such subvarieties are said to be _invariant_. We fix X and V for our discussion of several important special cases where maps may be shown to be equivariant.

Theorem 1.1: Given $f : X \longrightarrow Y$ a surjective proper map with connected fibers then f is equivariant, provided Y is normal, or f is smooth.

Proof: The hypotheses on f guarantee that $f_*(\mathcal{O}_X) \simeq \mathcal{O}_Y$. In fact the injectivity of the natural map follows from the surjectivity of f. Surjectivity follows by observing that a function g on X holomorphic on a neighborhood of the fibre is necessarily constant on the compact connected fibres and hence defines a continuous function $\bar{g}$ on Y. This function is holomorphic on any Zariski open set U where f admits a local section. If Y is normal, $\bar{g}$ is holomorphic since it is clearly weakly holomorphic. Since derivations of $\mathcal{O}_X$ define derivations of $f_*\mathcal{O}_X = \mathcal{O}_Y$, we see that vector fields descend.

As a particular corollary of the preceding, if $X \longrightarrow Y$ is a fibre bundle with fibre $\mathbb{P}^n$, then every vector field on X descends to Y. Conversely, given a vector bundle of rank $n+1$, $E \longrightarrow Y$ with $X \longrightarrow Y$ the associated projective bundle, the problem of lifting vector fields W from Y to X may be solved as follows. Cover Y by open sets U^α over which $E \xrightarrow{\sim} \mathbb{C}^n \times U^\alpha$. Let $x_0^\alpha \dots x_n^\alpha$ be the corresponding local dual basis of E^*. Holomorphic vector fields on $\mathbb{P}^n$ lift to certain fields on $\mathbb{C}^n$, namely the derivations on $\mathbb{C}[x_0,\dots,x_n]$ which preserve the degree of all polynomials, i.e. $\Sigma\, c_{ij} x_j \frac{\partial}{\partial x_i}$, and are specificable by the induced C-linear map $x_i \longrightarrow \Sigma\, c_{ij} x_j$. Derivations of the form $c\, \Sigma\, x_i \frac{\partial}{\partial x_i}$ descend to the trivial field on $\mathbb{P}^n$ (Euler's formula). Noting that $\Theta_{\mathbb{P}^n \times U^\alpha} = \Theta_{\mathbb{P}^n} \oplus \Theta_{U^\alpha}$, one sees that derivations on $\mathbb{P}^n \times U^\alpha$ lifting W are of the form $(g, x_i^\alpha) \longrightarrow W(g) x_i^\alpha + g\, \Sigma\, f_{ij} x_j^\alpha$ where $g, f_{ij} \in \Gamma(U^\alpha, \mathcal{O})$, i.e. the derivations are specified by giving a $\mathbb{C}$-linear $D_\alpha : \Gamma(U^\alpha, E^*) \longrightarrow \Gamma(U^\alpha, E^*)$, which satisfies $D_\alpha(g\sigma) = W(g)\,\sigma + g\, D_\alpha(\sigma)$ for $g \in \Gamma(U^\alpha, \mathcal{O})$, $\sigma \in \Gamma(U^\alpha, E^*)$, with D_α and D'_α giving the same derivation on $\mathbb{P}^n \times U^\alpha$ if and only if $(D_\alpha - D'_\alpha)(x_i^\alpha) = f\, x_i^\alpha$ for some $f \in \Gamma(U^\alpha, \mathcal{O})$. Note that in general the difference of two derivations D_α, D'_α both lifting W is determined by a linear map $E^* \longrightarrow E^*$. Given a collection of derivations D_α on U_α lifting W, they yield a global lift of W if and only if $D_\beta - D_\alpha : E^* \longrightarrow E^*$ is an $\mathcal{O}_{U^\alpha \cap U^\beta}$ multiple of the identity. The existence of such a global lifting is obstructed precisely by $D_\beta - D_\alpha \in H^1(Y, \mathrm{Hom}(E^*, E^*)/\mathcal{O}_Y)$. Recall the Atiyah-Chern class [1] $C(E^*) \in H^1(Y, \mathrm{Hom}(E^*, E^*) \otimes \Omega^1)$ which obstructs the construction of a $\tilde{D} : E^* \longrightarrow \Omega^1 \otimes E^*$ satisfying $\tilde{D}(f\sigma) = df \otimes \sigma + f\,\tilde{D}(\sigma)$. Given such a $\tilde{D}$, one may define $D : E^* \longrightarrow E^*$ by $D = i(W) \circ \tilde{D}$ where $i(W) : \Omega^1_Y \longrightarrow \mathcal{O}_Y$ is contraction and $D(f\sigma) = W(f)\,\sigma + f\,D(\sigma)$ is then a lift of W. Thus one obtains

<u>Theorem 1.2</u>: The vector field W on Y lifts **to** $\mathbb{P}(E) \longrightarrow Y$ if and only if $i(W)(C(E)) = 0 \in H^1(Y, \mathrm{Hom}(E^*, E^*)/\mathcal{O})$, where $C(E^*) \in H^1(Y, \mathrm{Hom}(E^*, E^*) \otimes \Omega^1)$ is the Atiyah Chern class.

The lifting is always possible provided E is a direct sum of line bundles, and $\mathrm{zero}(W) \neq \phi$. This follows from 1.5, and the additivity of $C(E)$. (Compare [10], pp. 109-110).

<u>Remark 1.3</u>: Note that the construction of a global $\mathbb{C}$-linear $D : E \longrightarrow E$ satisfying $D(f\sigma) = W(f)\,\sigma + f\,D(\sigma)$ is obstructed precisely by $i(W)C(E) \in H^1(Y, \mathrm{Hom}(E,E))$. The vanishing of this class is necessary and sufficient for W to lift to $E^* \longrightarrow Y$ in such a way that it descends to $\mathbb{P}(E^*) \longrightarrow Y$. The real importance of this obstruction lies in another direction. Given a "W-connection" $D : E \longrightarrow E$ as above assume moreover that the global sections of E span the fibre of E at at least one point (and hence actually on an open set U). The map $H^0(U,E) \times U \longrightarrow E_{|U}$ defines a varying $n+1$ quotient of the $N+1$ dimensional space $H^0(U,E)$, i.e. a holomorphic map $U \longrightarrow \mathrm{Grass}\,(n,N)$. The "W-connection" D gives rise to a global vector field on Grass (n,N), which is tangent to the image of U, and induces W on U. Hence the pair (E,D) defines an <u>equivariant</u> <u>rational</u> map $Y \longrightarrow \mathrm{Grass}\,(n,N)$. We consider explicitly the case where E is a line bundle. Letting $\sigma_0, \dots, \sigma_N$ be a basis for $H^0(Y,E)$ the rational map $Y \longrightarrow \mathbb{P}^N$ is defined by $y \longrightarrow (\sigma_0(y), \dots, \sigma_N(y))$. Thus the σ_i are the restrictions to Y of the

homogeneous coordinates x_i . D induces a $\mathbb{C}$-linear map $H^0(Y,E) \longrightarrow H^0(Y,E), \sigma_i \longrightarrow \Sigma c_{ij}\sigma_j$, which yields the vector field $\tilde{D} = \Sigma c_{ij} x_j \frac{\partial}{\partial x_i}$ on $\mathbb{P}^n$. The function x_j/x_i restricts to $f = \sigma_j/\sigma_i$ on Y , i.e. $\sigma_j = f\sigma_i$ so that $D(\sigma_j) = W(f)\sigma_i + f D(\sigma_i)$ or rewriting $\frac{D(\sigma_j)\sigma_i - \sigma_j D(\sigma_i)}{\sigma_i^2} = W(f)$. But the left hand term is $\tilde{D}(x_j/x_i)$ on Y , as asserted. Consequently,

Theorem 1.4: Given a line bundle $L \longrightarrow X$ and given $V \in H^0(X,\Theta)$, if $C(L) \in H^1(X,\Omega^1)$ is annihilated by $i(V) : H^1(X,\Omega^1) \longrightarrow H^1(X,\mathcal{O})$, then L defines an equivariant rational map $X \longrightarrow \mathbb{P}^N$, $N = \dim H^0(X,L)$. Conversely, given that L defines an equivariant map and that the map is globally defined then $i(V)\ (C(L)) = 0$.

Proof: The first assertion is proven above. Conversely, if $f : X \longrightarrow \mathbb{P}^N$ is globally defined with $f_*(V) = V'$, then, since the sheaf of sections $\underline{L}$ of L is $f^*(\mathcal{O}(1))$, we see that $i(V)C(L) = f^*(i(V')\ C(\mathcal{O}(1))) = 0$, since $H^1\ (\mathbb{P}^N,\mathcal{O}) = 0$.

Characteristic zero hypotheses were unnecessary for the preceding results. The next theorem, which gives a simple criterion for $i(V)\ C(L) = 0$, requires in its proof the "Hard Lefschetz" result $\mathbb{L}^{n-1} : H^1(X,\mathcal{O}) \xrightarrow{\sim} H^n(X,\Omega^{n-1})$ (for X a Kahler n-manifold, and $\mathbb{L} \in H^1(X,\Omega^1)$ the fundamental class).

Theorem 1.5: (Generalizes Matsushima [12], Lichnerowicz [11]): Given a compact Kahler n-manifold X and a $V \in H^0(X,\Theta)$, the following properties are all equivalent:

(a) $\operatorname{zero}(V) \neq \emptyset$

(b) $i(V) : H^0(X,\Omega^1) \longrightarrow H^0(X,\mathcal{O})$ is zero

(c) $i(V) : H^n(X,\Omega^n) \longrightarrow H^n(X,\Omega^{n-1})$ is zero

(d) $i(V) : H^1(X,\Omega^1) \longrightarrow H^1(X,\mathcal{O})$ is zero

(e) $i(V)(\mathbb{L}) = 0$ where $\mathbb{L} \in H^1(X,\Omega^1)$ is the Kahler class

Proof: $a \longrightarrow b$ for $\phi \in H^0(\Omega^1)$, $i(V)(\phi)$ is a global holomorphic function, hence constant, and vanishes on zero (V).

$b \longrightarrow c$ Serre duality

$d \longrightarrow e$ trivial

$e \longrightarrow a$ If $i(V)(\mathbb{L}) = 0$ then also denoting by $\mathbb{L}$ a closed (1,1) form representing $\mathbb{L}$, $i(V)(\mathbb{L}) = \bar{\partial} f$ for some function f . Hence $(i(V) + \bar{\partial})(\mathbb{L} - f) = 0$, and $(i(V) + \bar{\partial})((\mathbb{L} - f)^n = 0$. Consider the double complex $E^{p,q}$ of global C^∞ forms with total differential $i(V) + \bar{\partial}$, where $i(V) : E^{p,q} \longrightarrow E^{p-1,q}$ and $\bar{\partial} : E^{p,q} \longrightarrow E^{p,q+1}$. The form $(\mathbb{L} - f)^n$ is a cycle. If $\operatorname{zero}(V) = \emptyset$ then the rows $i(V) : \longrightarrow E^{p,q} \longrightarrow E^{p-1,q} \longrightarrow \ldots$ are exact (fixing $\lambda \in E^{1,0}$ such that $i(V)(\lambda) = 1$, wedge product with λ defines a homotopy for $i(V)$). It follows that $(\mathbb{L} - f)^n = i(V) + \bar{\partial})(\psi)$ for some ψ . But then necessarily $\mathbb{L}^n = \bar{\partial}(\psi_{n,n-1})$ and this contraducta $\mathbb{L}^n \neq 0 \in H^n(\Omega^n)$.

$c \longrightarrow d$ It suffices to show $c \longrightarrow e$, since c does not depend on the choice of Kahler class $\mathbb{L}$, and any element of $H^1(X,\Omega^1)$ is a linear combination of Kahler classes. Now $\mathbb{L}^n \in H^n(\Omega^n)$ hence $0 = i(V)\,\mathbb{L}^n = n\,\mathbb{L}^{n-1}(i(V)\,\mathbb{L})$. Since $\mathbb{L}^{n-1}$: $H^1(X,\mathcal{O}) \xrightarrow{\sim} H^n(X,\Omega^{n-1})$, $0 = i(V)(\mathbb{L})$ as required.

It was shown in [5] that zero$(V) \neq \emptyset$ if and only iff $i(V) : H^p(\Omega^q) \longrightarrow H^p(\Omega^{q-1})$ is zero for all values of p and q.

For any complex space X, let $\Theta_0(X)$ denote the set of vector fields V with zero$(V) \neq \emptyset$.

Corollary 1.6: If $H^1(X,\mathcal{O}) = 0$, then $H^0(X,\Theta) = \Theta_0$, for X a Kahler manifold.

Corollary 1.7: (Borel-Sommese [16]). Given X a Kahler manifold, $V \in \Theta_0(X)$ and Y an invariant subvariety, then $V_{|Y} \in \Theta_0(Y)$.

Proof: One may assume Y nonsingular since Sing(Y), Sing(Sing(Y)), are invariant and the last stratum is nonsingular. Under $f : Y \longrightarrow X$ one has $f^*(\mathbb{L})$ is a Kahler class on Y and the result follows by a $\longrightarrow$ d on X and d $\longrightarrow$ a on Y.

Theorem 1.8: [generalizes Blanchard [2]). If X is a Kahler manifold, or a normal projective variety and $L \longrightarrow X$ is a line bundle then the rational map $X \longrightarrow \mathbb{P}(H^0(L)^*)$ is Θ_0 equivariant.

Proof: For X a manifold this follows from 1.4, 1.5. For X normal, let $f : \hat{X} \longrightarrow$ be an equivariant resolution of X (i.e. all vector fields lift, cf. 1.12 below). For any $V \in \Theta_0$ its lift $\hat{V}$ will be in $\Theta_0(\hat{X})$ (cf. lemma 1.13, below). Then the commutative diagram

$$\begin{array}{ccc} H^1(\Omega^1_{\hat{X}}) & \xrightarrow{i(\hat{V})} & H^1(\mathcal{O}_{\hat{X}}) \\ {\scriptstyle f^*}\uparrow & & \uparrow{\scriptstyle f^*} \\ H^1(\Omega^1_X) & \xrightarrow{i(V)} & H^1(\mathcal{O}_X) \end{array}$$

completes the proof once one observes that $i(\hat{V}) = 0$, by 1.5 and that $f^* : H^1(\mathcal{O}_X) \longrightarrow H^1(\mathcal{O}_{\hat{X}})$ is injective. This latter follows from the Leray spectral sequence, and the remark that $f_*(\mathcal{O}_{\hat{X}}) = \mathcal{O}_X$ since X is normal.

The theorem may fail if X is not normal, for example if X is the nodal curve obtained by identifying 0 and ∞ in $\mathbb{P}^1$, $\Theta_0 = \mathbb{C}$ acts nontrivially on Pic(X). The main theorem in the nonnormal case is:

Theorem 1.9: (cf. Siedenberg [14]): Given X a complex space and $f : \tilde{X} \longrightarrow X$ its normalization. Then all vector fields on X lift to $\tilde{X}$ and Θ_0 lifts to $\tilde{\Theta}_0$.

Proof: Characteristic zero is essential for the theorem [14]. The simplest proof is to observe that a vector field V on X integrates to give a 1-parameter family of local automorphisms of X. Since $\tilde{X}$ is functorial in X, the family lifts to a 1-parameter family on $\tilde{X}$. Moreover zeroes of V correspond to fixed points of the flow. The fibre over any fixed point is mapped to itself. The fibre is therefore pointwise fixed since it is discrete and the maps are homotopic to the identity.

Theorem 1.10: Let X be a complex space, $V \in H^0(X,\Theta)$ and let $Z \subset X$ be a V-invariant subvariety. If $\pi: \hat{X} \longrightarrow X$ is the monoidal transform centered at Z, then V lifts uniquely to $\hat{V} \in H^0(\hat{X},\Theta)$. If $V \in \Theta_0(X)$ then $\hat{V} \in \Theta_0(\hat{X})$.

Proof: Viewing V as a derivation $V: \mathcal{O}_X \longrightarrow \mathcal{O}_X$ and letting $I_Z \subseteq \mathcal{O}_X$ be the sheaf of ideals defining Z, we may express the invariance of Z by $V(I_Z) \subseteq I_Z$. Consequently $V(I_Z^n) \subseteq I_Z^n$ for all n. Hence V gives rise uniquely to grading preserving derivation of the sheaf of rings $A = \mathcal{O}_X \oplus I \oplus I^2 \oplus \ldots\ldots$ Since $\hat{X} = \mathrm{Proj}(A)$ the result follows. For a more explicit local construction of $\hat{V}$, useful in the sequel, proceed as follows. Let $f_0,\ldots,f_r$ be local generators for I_Z over $U \subseteq X$. Then $\hat{U} = \pi^{-1}(U)$ is the Zariski closure in $U \times \mathbb{P}^r$ of the subvariety of $(U-Z)\times\mathbb{P}^r$ defined by $X_i f_j - X_j f_i = 0$ where $X_0,\ldots,X_r$ are homogeneous coordinates on $\mathbb{P}^r$. Since Z is invariant, $V(f_i) = \Sigma a_{ij} f_j$ for suitable functions a_{ij}. Consider the vector field $\hat{V} = V + \Sigma a_{\ell k} X_k \frac{\partial}{\partial X_\ell}$ on $U \times \mathbb{P}^r$. $\hat{V}$ is tangent to $\hat{U}$ since

$$\hat{V}(X_i f_j - X_j f_i) = X_i V(f_j) - X_j V(f_i) + \sum_k a_{ik} X_k f_j - \sum_k a_{jk} X_k f_i$$

$$= \sum_k a_{jk} f_k X_i - \sum_k a_{ik} f_k X_j + \sum_k a_{ik} X_k f_j - \sum_k a_{jk} X_k f_i$$

$$= \sum_k a_{jk} (f_k X_i - X_k f_i) + \sum a_{ik} (X_k f_j - f_k X_j).$$

The restriction of $\hat{V}$ to $\hat{U}$ is the required lift.

If $x \in U$ is a zero of V, the vector field $\hat{V}$ on $U \times \mathbb{P}^r$ is tangent to $\{x\} \times \mathbb{P}^r$ and to the subvariety $\hat{U} \cap \{x\} \times \mathbb{P}^r$. Moreover $\hat{V}$ has a zero on $\hat{U} \cap \{X\} \times \mathbb{P}^r$ by 1.7.

Remark: Given a vector field $W = \Sigma c_{ij} X_i \frac{\partial}{\partial X_j}$ on $\mathbb{P}^r$, its zeroes occur precisely at points of C^{r+1} where W is proportional to the vector field $\Sigma X_j \frac{\partial}{\partial X_j}$, i.e. the points $(a_0,\ldots,a_r)$ for which $\sum_{ij} c_{ij} a_i \frac{\partial}{\partial X_j} = \lambda \sum_j a_j \frac{\partial}{\partial X_j}$ for suitable λ, that is $(a_0,\ldots,a_r)$ is an eigen-vector for (c_{ij}). The zeroes are isolated if and only if (c_{ij}) has precisely one eigen-vector for each eigen-value, e.g. if there are $r+1$ distinct eigen-values. In particular in the preceding proof we noted that at a zero $x \in X$ of V the induced vector field on $\mathbb{P}^r \times \{x\}$ is given by the matrix $a_{ij}(x)$. The vector field on $\hat{U}$ will have only isolated zeroes over x if x itself is isolated and if moreover the matrix $a_{ij}(x)$ has distinct eigen-values. This may be given an intrinsic interpretation, namely choose $f_0,\ldots,f_r$ so that their images give a basis for $I/M_x \cdot I$ (where I is the ideal of Z and M_x the ideal functions vanishing at x). Since V vanishes at x, $V(\mathcal{O}_x) \subseteq M_x$ and it follows readily that $V(M_x^n) \subseteq M_x^n$. In particular V defines a natural linear transformation $L(V): M_x/M_x^2 \longrightarrow M_x/M_x^2$. The restriction of this transformation to $I/M_x \cdot I$ has the matrix $a_{ij}(x)$.. Thus, if the eigen-values of $L(V)$ are distinct, those of $a_{ij}(x)$ are. Moreover, x is an isolated fixed point itself, if $L(V)$ has no zero eigen-value. Indeed, if x is not isolated, and W is locus of zeroes of V then $L(V)$ is zero on the cotangent space to x in W.

In the sequel we shall need to ensure that an isolated zero remains isolated under successive monoidal transforms. We have seen that given a monoidal transform $\hat{X} \longrightarrow X$ then we may ensure that over a zero x of V the zeroes of its lift $\hat{V}$ will be isolated by assuming that $L(V)$ has distinct non zero eigen-values. We seek to determine the eigen-values of $L(\hat{V})$ at one of these isolated zeroes. One may assume $f_0, \ldots, f_r$ chosen so that their images in $I/M_x \cdot I$ form a basis of eigen-vectors and extend this basis by $f_{r+1} \ldots f_n$ yielding an eigen-basis for M_x/M_x^2 with corresponding eigen-values $\lambda_0, \ldots, \lambda_n$. We investigate the eigen-values at $x = (1,0,\ldots,0) \in \hat{X}^1$. At this point the functions $f_0, f_{r+1}, \ldots, f_n$ and $x_i = X_i/X_0$ generate M/M^2 and $f_i = x_i f_0 \in M^2$, for $i = 1, \ldots, r$. Note that

$$\hat{V}(x_i) = (\hat{V}(X_i)X_0 - \hat{V}(X_0)X_i)/X_0^2 = \lambda_i X_i X_0^{-1} - \lambda_0 X_i X_0^{-1} = (\lambda_i - \lambda_0) x_i .$$

Hence the eigen-values are $(\lambda_0, \lambda_{r+1}, \ldots, \lambda_n, \lambda_1 - \lambda_0, \ldots, \lambda_r - \lambda_0)$. Note that if the eigen-values $\lambda_0, \ldots, \lambda_n$ are linearly independent over $\mathbb{Z}$ then the resulting eigen-values are linearly independent (and, in particular, distinct). Moreover, under any succession of monoidal transforms this independence (and hence distinctness) will be preserved. This motivates

Definition: A zero $x \in X$ of the vector field V will be called generic if the eigen-values of $L(V): M_x/M_x^2 \longrightarrow M_x/M_x^2$ are linearly independent over $\mathbb{Z}$

We have observed

Proposition 1.11: If $x \in X$ is a generic zero of V, and if $\pi: \hat{X} \longrightarrow X$ is a composition of equivariant monoidal transformations then every zero of $\hat{V}$ on $\pi^{-1}(x)$ is isolated, and generic.

This notion of generic zero goes back essentially to Poincaré [13] and the following remark is derived from his thesis.

Remark: Let $x \in X$ be a generic zero of $V \in H^0(X,\Theta)$; then X is nonsingular at x and one may choose "formal" local coordinates x_i, $i = 1, \ldots, n$ at x so that $V = \Sigma \lambda_i x_i \frac{\partial}{\partial x_i}$, where λ_i are the eigen-values of $L(V)$ at x.

Any surface may be resolved by a succession of (a) normalizations and (b) monoidal transformations at isolated singular points. Consequently, every surface X (char. 0) has an equivariant resolution $\pi: \hat{X} \longrightarrow X$, (i.e. all vector fields lift from X to $\hat{X}$, cf. [3]). This follows immediately from Theorems 1.9, 1.10 and the remark that in char. 0 every isolated singular point is automatically invariant (a fact which is readily seen to be false in general for char. p). This result has been vastly generalized by the following, soon to be published, result of Hironaka, cf. [8].

Theorem 1.12: Given a compact complex space X, there exists a sequence $\hat{X} = X_n \longrightarrow X_{n-1} \longrightarrow \cdots \longrightarrow X_1 \longrightarrow X_0 = X$ with $\hat{X}$ nonsingular and $X_{i+1} \longrightarrow X_i$ a monoidal transformation with nonsingular center Z_i which is invariant under all vector fields V on X_i which descend to $X_0 = X$. Hence $\hat{X} \longrightarrow X$ is equivariant with respect to $H^0(X,\Theta)$.

Remark: If X is normal then every vector field on $\hat{X}$ (and on any of the X_j) descends to X (Theorem 1.1). In this case the monoidal transformations $X_{i+1} \longrightarrow X_i$ are strictly equivariant, i.e. $H^0(X_{i+1},\Theta) \xrightarrow{\sim} H^0(X_i,\Theta)$. If X is not normal, it is in general impossible to resolve in a strictly equivariant manner.

Lemma 1.13: Let $\hat{X} \longrightarrow X$ be an equivariant resolution of singularities. If $V \in \Theta_0(X)$ then $\hat{V} \in \Theta_0(\hat{X})$, in fact $\hat{V}$ has a zero lying over every zero of V . If V has a generic zero, so does $\hat{V}$.

Proof: The result follows from 1.10, 1.11, 1.12.

In our initial study of generic zeroes we employed the following

Conjecture: Given $V \in H^0(X,\Theta)$ and $x \in X$ a generic zero of V , let $f : X \longrightarrow Y$ be a surjective, V-equivariant map with $y = f(x)$ a simple point of Y . Then y is a generic zero of $f_*(V)$.

This would follow from the assertion that if $\mathcal{O}_y \longrightarrow \mathcal{O}_x$ is an injective map of convergent power series rings, then the induced mapping on formal power series is also injective. Counterexamples to this general assertion have been obtained by Gabrielov.

To avoid this difficulty we introduce a larger class of vector field.

Definition: Given $V \in H^0(X,\Theta)$, a point $x \in X$ will be said to be substantiating for V if for any $f, g \in \mathcal{O}_x$, the vanishing of the Wronskian $W(f,g) = V(f)g - fV(g)$ implies that f and g are $\mathbb{C}$-linearly dependent. V will be called substantial if it has a substantiating point.

Note that on a complex analytic manifold X , if $V(x) \neq 0$ then x is never substantiating (unless $\dim X = 1$ in which case all points are substantiating if $V \neq 0$). An easy calculation shows that if x is a generic zero of V , then it is substantiating.

Lemma 1.14: Given an algebraic variety X and $V \in H^0(X,\Theta)$, if V is substantial then every point of X is substantiating, i.e. V is substantial if and only if it induces a substantial derivation of the field of global meromorphic functions.

Proof: Assume x is substantial for V . Consider the derivation V induced on $k(X)$ = the field of global meromorphic functions. Given $f, g \in k(X)$ choose $a,b,c \in \mathcal{O}_x$ such that $f = \frac{a}{c}$, $g = \frac{b}{c}$. Note that $W(f,g) = \frac{1}{c^2} W(a,b)$, hence if $0 = W(f,g)$ it follows that a and b are $\mathbb{C}$-linearly dependent (and therefore so are f and g). The result follows by noting that every local ring is a subring of $k(X)$.

Corollary 1.15: Suppose given $V \in H^0(X,\Theta)$ and $f : X \longrightarrow Y$ a rational V-equivariant map with dense image. If V is substantial on X, then $f_*(V)$ is substantial on Y . If f is birational, then V is substantial if and only if $f_*(V)$ is.

Proof: f induces an injection $k(Y) \longrightarrow k(X)$, which is an isomorphism if f

is birational.

Note that in general the algebraic local ring at a point x is a subring of the analytic local ring. Hence any algebraic vector field which is analytically substantial at x, is algebraically substantial. The converse is false, as may be seen by regarding a vector field on an n-torus whose integral curves are dense. This field is algebraically substantial, but for $n \geq 2$ is not analytically substantial.

Problem: If V is an analytic vector field substantial at x, and $V(x) = 0$, does it follow that x is an isolated zero of V ?

The following remarks about derivations of $k(X)$ are not used in the sequel, but serve to clarify the notion of substantiality. Given $D : k(X) \longrightarrow k(X)$ a $\mathbb{C}$-linear derivation, denote by G the set of eigen-values of D, i.e. $\lambda \in \mathbb{C}$ such that $D(f) = \lambda f$ for some $f \neq 0$. One shows readily that G is a free abelian group of rank $\leq$ tr deg $(k(X))$ where $N \subseteq k(X)$ is the subfield of D-invariants. In particular $\mathrm{rank}(G) \leq \dim(X)$, and equality can hold only if D is substantial. (The converse is in general false, i.e. $\frac{d}{dx}$ on $C(x)$ is substantial and has $G = \{0\}$, or more pathologically consider the derivation given by a skew vector field on the torus). Note that $k(X)$ admits substantial derivations with $\mathrm{rank}(G) = \dim(X)$, obtained by making the elements of a separating transcendence basis the eigen-functions and extending uniquely to $k(X)$.

Our results imply that if $k(X)$ has a nonsingular projective model for which such a D is holomorphic, and vanishes somewhere, then $k(X)$ is a purely transcendental extension of $\mathbb{C}$.

§2. Remarks on substantial vector fields.

(a) Examples of substantial vector fields.

Fixing $\lambda_0, \ldots, \lambda_n \in \mathbb{C}$ which are linearly independent over $\mathbb{Z}$, the vector field $\Sigma \lambda_i X_i \frac{\partial}{\partial X_i}$ is substantial on $\mathbb{P}^n$ (in fact its zeroes are all generic). Similarly, consider the variety $\mathrm{Grass}(r,N)$ of r planes in $\mathbb{C}^N$. The r planes near a given r plane P may be coordinatized by the $(r \times N - r)$ matrices $((a_{ij}))$ where $v_1, \ldots, v_N$ give a basis for $\mathbb{C}^N$ with the first r giving a basis for P and $v_i + \sum_{j=1}^{N-r} a_{ij} v_{j+r}$ give a basis for the plane corresponding to $((a_{ij}))$. Picking independent λ_{ij}, one may extend the one parameter family $a_{ij} \longrightarrow e^{t\lambda_{ij}} a_{ij}$ to a one parameter family of collineations of the Plucker coordinate space $\mathbb{P}^{\binom{N}{r}-1}$ which leave $\mathrm{Grass}(r,N)$ invariant. The corresponding vector field on $\mathrm{Grass}(r,N)$ is substantial (having again, generic zeroes). Similar constructions may be employed for flag manifolds. Moreover, if X,V (resp. Y,W) are two spaces having vector fields with generic zeroes at x, (resp. y) then $(V, \gamma \cdot S)$ will have a generic zero at $(x,y) \in X \times Y$ provided $\gamma \in \mathbb{C}$ is sufficiently general. Hence, the standard homogeneous rational varieties all admit substantial (even generic) vector fields.

(b) Classification of all substantial vector fields on $\mathbb{P}^n$.

Consider the global vector field $V = \Sigma\, a_{ij} X_j \frac{\partial}{\partial X_i}$ on $\mathbb{P}^n$. By a linear change of coordinates, we may assume that the matrix $A = ((a_{ij}))$ is in Jordan normal form with the largest Jordan block being in the upper-left corner. Moreover, subtracting a multiple of $\Sigma\, X_i \frac{\partial}{\partial X_i}$, one may assume that the eigen-value for this Jordan block is zero. Let $\lambda_1,\ldots,\lambda_k$ be the eigen-values corresponding to the other Jordan blocks.

<u>Proposition 2.1</u>: Such a vector field $V \in H^0(P^n,\Theta)$ is substantial if and only if the largest Jordan block is at most 2×2, the Jordan blocks corresponding to the λ_i are 1×1 and the λ_i are linearly independent over $\mathbb{Z}$ (in particular each eigen-value has only one Jordan block).

<u>Proof</u>: Assume V is substantial. Suppose the largest Jordan block is $(r+1)\times(r+1)$. We have assumed that such a block corresponds to the variables $X_0,\ldots,X_r$ and has eigen-value 0. Hence $V(X_r) = 0$, $V(X_{r-1}) = X_r$ and $V(X_{r-2}) = X_{r-1}$. Letting $x_i = X_i/X_r$, we see that $V(x_{r-1}) = 1$ and $V(x_{r-2}) = x_{r-1}$, whence $V(2x_{r-2} - x_{r-1}^2) = 0$. But $2x_{r-2} - x_{r-1}^2$ is clearly a non-constant meromorphic function on $\mathbb{P}^n$, if $r \geq 2$. Since V is substantial, $r \leq 1$. Assuming $r = 1$, we let $y = X_0/X_1$ and note that $V(y) = 1$. Suppose there is another 2×2 Jordan block, i.e. $V(X_{i+1}) = \lambda X_{i+1}$ and $V(X_i) = \lambda X_i + X_{i+1}$. Then $V(X_i/X_{i+1}) = 1$. Hence $V(y - X_i/X_{i+1}) = 0$ and V is not substantial. Thus there is at most one 2×2 block. Let $x_j = X_{j+1}/X_1$, so that $V(x_j) = \lambda_j x_j$. Note that if $I = (i_1,\ldots,i_k)$ then $V(x^I) = (I\cdot\lambda)x^I$ where $I\cdot\lambda = \Sigma\, i_j\lambda_j$. If the λ_j are $\mathbb{Z}$-dependent, i.e. for some I, $I\cdot\lambda = 0$, then $V(x^I) = 0$ contradicting the substantiality of V.

Conversely, assuming at most one 2×2 Jordan block and $\mathbb{Z}$-independence of the λ_i we verify substantiality by induction on n. For $n = 1$ all vector fields are substantial.

Assume there is a 2×2 Jordan block, and fixing y, x_i, satisfying $V(y) = 1$, $V(x_i) = \lambda_i x_i$, as above, we verify substantiality of V on $R = \mathbb{C}[y,x_1,\ldots,x_{n-1}]$. (The case of 1×1 Jordan blocks is handled below).

Assume $0 \neq F$, $G \in R$ and $W(F,G) = 0$. Expand as polynomials $F = \Sigma\, f_i y^i$, $G = \Sigma\, g_j y^j$ where $f_i, g_j \in \mathbb{C}[x_1,\ldots,x_k]$ and let r (resp. s) be the y degree of F (resp. G). We may assume $r \geq s$ and obtain strict inequality by subtracting a constant multiple of F from G. We must show $G = 0$. Now

$$0 = W(F,G) = \sum_{i,j} W(f_i,g_j)\, y^{i+j} + \sum_{i,j} (i-j)\, f_i\, g_j\, y^{1+j-1}$$

Therefore $W(f_r,g_s) = 0$, (it is the highest coefficient). But then $g_s = k f_r$ (by induction) and clearly $k \neq 0$, assuming $G \neq 0$. The coefficient of y^{r+s-1}

$$0 = W(f_r,g_{s-1}) + W(f_{r-1},g_s) + (r-s)f_r g_s$$

Noting that $W(f_{r-1},g_s) = W(f_{r-1},kf_r) = -W(f_r,kf_{r-1})$ and $W(f_r,g_{s-1}) - W(f_r,kf_{r-1}) = W(f_r,g_{s-1} - kf_{r-1}) = -W(g_{s-1} - kf_{r-1}, f_r)$, we may rewrite the equation as

$$k(r-s)f_r^2 = W(g_{s-1} - kf_{r-1}, f_r) .$$

However, we will show that the equation

$$(*) \qquad \eta b^2 = W(a,b)$$

has no solution with $a,b \in C[x_1,\dots,x_k]$, $\eta \in \mathbf{C}$ and $b,\eta \neq 0$. It follows that $G = 0$ in the preceding.

Let $a = \Sigma a_I x^I$ and $b = \Sigma b_J x^J$ so that $W(a,b) = \Sigma(I-J)\cdot\lambda\, a_I b_J x^{I+J}$. We lexicographically order the multi-indices $I = (i_1,\dots,i_k)$ by $I < I'$ if $i_k = i'_k$ for $k < t$ and $i_t < i'_t$. Note that if $I < I'$ and $J \leq J'$ then $I + J < I' + J'$. The leading coefficient ρ_{I_0} of a polynomial $\Sigma \rho_I x^I$ is that $\rho_I \neq 0$ for which I is minimal.

Let a_{I_0} (resp. b_{J_0}) be the leading coefficient of a (resp. b). The leading coefficient of $W(a,b) - \eta b^2$ is either $\eta b_{J_0}^2$ if $J_0 \leq I_0$, or $(I_0 - J_0)\cdot\lambda\, a_{I_0} b_{J_0}$ if $J_0 > I_0$. If $\eta \neq 0$ then this leading coefficient is clearly nonzero (recalling the $\mathbf{Z}$-independence of λ_i), and the equation (*) cannot be solved.

Note moreover that the only solutions of $W(a,b) = 0$ are of the form $b = ka$. Indeed, the leading coefficient of $W(a,b)$ will be $(I_0 - J_0)\cdot\lambda\, a_{I_0} b_{J_0}$ unless $I_0 = J_0$. However, by subtracting a constant multiple ka from b one may assume $I_0 \neq J_0$, and thus one concludes that $W(a,b) = 0$ implies $b = ka$. This last remark yields the proof for the case where all Jordan blocks are 1×1 and completes the proof of the proposition.

The following discussion of the homogeneous eigen-functions of V acting on $\mathbf{C}[X_0,\dots,X_n]$ is useful in the sequel. We assume V is substantial.

<u>Case 1</u>: $V = \Sigma \lambda_i X_i \frac{\partial}{\partial X_i}$ with $\lambda_0 = 0$ and $\lambda_1, \dots, \lambda_n$ $\mathbf{Z}$-independent. The homogeneous eigen-functions are precisely the monomials. X^I has eigen-value $I\cdot\lambda$. Indeed, let $F(X_0,\dots,X_n) \neq 0$ be homogeneous of degree d with $V(F) = \eta F$. Let $f(x_1,\dots,x_n) = F/X_0^d$. Then $V(f) = \eta f$. Expanding $f = \Sigma f_I x^I$, then letting $\lambda = (\lambda_1,\dots,\lambda_n)$ we have

$$V(f) = \Sigma f_I(I\cdot\lambda)\, x^I = \Sigma \eta f_I x^I$$

and clearly $(\eta - (I\cdot\lambda))f_I = 0$. If $f_I \neq 0$, $\eta = I\cdot\lambda$. Since $I\cdot\lambda \neq J\cdot\lambda$ for $I \neq J$ there is at most one $f_I \neq 0$.

<u>Case 2</u>: $V = \Sigma_i \lambda_i X_i \frac{\partial}{\partial X_i} + X_1 \frac{\partial}{\partial X_0}$ with $\lambda_0 = \lambda_1 = 0$ and $\lambda_2, \dots, \lambda_n$ $\mathbf{Z}$-independent. The homogeneous eigen-functions are precisely the monomials X^I, where $I = (0, i_1, \dots, i_n)$ and $I\cdot\lambda$ is the eigen-value.

Indeed, letting $y = X_0/X_1$ and $x_i = X_i/X_1$ $i = 2, \dots, n$, then given $F(X_0, \dots, X_n)$ of degree d with $V(F) = \eta F$ form $f(y, x_2, \dots, x_n) = F/X_1^d$; $V(f) = \eta f$. Expand $f = \Sigma f_i y^i$ where $f_i \in C[x_2, \dots, x_n]$. Comparing coefficients in

$$\Sigma V(f_i) y^i + \Sigma i f_i y^{i-1} = \Sigma \eta f_i y^i$$

one finds $V(f_r) = \eta f_r$ where r is the y degree of f. By case 1, $f_r = c_r x^I$ for some $I = (i_2, \ldots, i_n)$ and $\eta = I \cdot \lambda$.

Examining the coefficient of X^I in the coefficients for y^{r-1} in our equation, one finds

$$(I\cdot\lambda)c_{r-1} + r\, c_r = (I\cdot\lambda)\, c_{r-1}$$

and necessarily $r = 0$, i.e. f is independent of y. Thus

Remark 2.2: If V is substantial, the only homogeneous eigen-functions are monomials. Moreover, the group of eigen-values of V on $k(\mathbb{P}^n)$ is the free group with basis the nonzero eigen-values of the matrix of V.

(c) Examples of nonrational surfaces

Let X be a projective algebraic variety of dimension 2, and let $V \in H^0(X,\Theta)$ have only isolated zeroes. By successively normalizing and blowing up singular points of X one obtains $f : \hat{X} \longrightarrow X$ and $\hat{V} \in H^0(\hat{X},\Theta)$ with $f_*(\hat{V}) = V$, and with $\hat{X}$ a nonsingular projective algebraic variety. Moreover, $\hat{V}$ has (not necessarily isolated) zeroes on $\hat{X}$, cf. 1.13. Note that $\hat{X}$ is birational to a ruled surface since all plurigenera of $\hat{X}$ vanish ([10]). This may be seen more directly. Assuming $\hat{X}$ is not rational, we know the zeroes of V are not isolated and moreover that $\dim(\mathrm{Alb}(\hat{X})) = \dim H^0(X,\Omega^1) \neq 0$ (since otherwise the vanishing of P_2 implies rationality). The image C of $\hat{X} \longrightarrow \mathrm{Alb}(\hat{X})$ is necessarily 1 dimensional, since $H^0(\hat{X},\Omega^2) = 0$. C is a nonsingular curve of genus ≥ 1 (cf. [15] p. 54).

Since $\hat{V}$ has zeroes, $(\hat{V},\phi) = 0$ for all $\phi \in H^0(X,\Omega^1)$ so that $f : \hat{X} \longrightarrow \mathrm{Alb}(\hat{X})$ is equivariant with $f_*(\hat{V}) = 0$. Thus $\hat{V}$ is tangent to all the fibres of f, and hence by Borel-Sommese has a zero on each component of each fibre (cf. [16]). $\hat{V}$ cannot vanish identically on any fibre, since a fibre has self intersection zero, hence is not collapsed under $\hat{X} \longrightarrow X$ and on X the zeroes of V are isolated. Similarly, if $\hat{X}$ is chosen to be a minimal resolution, $\hat{V}$ will not vanish on any component of any fibre. Since the components of the fibre are curves admitting a nontrivial vector field with isolated zeroes they are all rational, i.e. $\hat{X}$ is clearly ruled and the zeroes of $\hat{V}$ are curves transverse to the fibres.

One can easily make examples of such $\hat{X}$ and X. Namely let $Y = \mathbb{P}^1 \times C$, where $g(C) \geq 1$. Consider the vector field $W = (\begin{pmatrix} 0 & 0 \\ 0 & \lambda \end{pmatrix}, 0)$ with $\lambda \neq 0$, which is "tangent to the fibres". The zeroes of W are $C_1 = (1,0) \times C$ and $C_2 = (0,1) \times C$. These varieties have self intersection zero. Let $\hat{X}$ be obtained from Y by blowing up one point on each C_i, and let $\hat{V}$ be the lift of W. The zeroes of $\hat{V}$ are the proper transforms $\hat{C}_i$ of C_i and two other isolated points. Since $(\hat{C}_i,\hat{C}_i) = -1$ one may blow down the $\hat{C}_i$ to obtain X and V.

§3. Equivariant projections. The induction step.

In the following sections we prove by induction on $n = \dim X$ that if X is a complex projective algebraic variety and $V \in H^0(X,\Theta)$ is substantial on X and zero $(V) \neq \emptyset$, then there is an equivariant birational map $X \longrightarrow \mathbb{P}^n$.

Remarks: The assumption $\mathrm{zero}(V) \neq \emptyset$ is necessary since any vector field W on $\mathbb{P}^n$ has zeroes, and hence given $\mathbb{P}^n \longrightarrow X$ equivariant, rational one employs Hironaka's results to find $Y \longrightarrow \mathbb{P}^n$ by a sequence of equivariant monoidal transformations so that $f : Y \longrightarrow X$ is equivariant and holomorphic. The lift $\hat{W}$ of W to Y has zeroes (1.13) and hence $f_*(\hat{W})$ has zeroes on X. Moreover, the hypothesis is needed to rule out various examples, e.g. skew vector fields on abelian varieties. The hypothesis that V be holomorphic is immaterial, but is employed to eliminate a multitude of counter-examples e.g. any X admits a meromorphic V satisfying the other hypotheses. A more detailed study of the admissible polar loci is currently being carried out. The projective hypothesis may be replaced by "complete" by noting that the standard proof of the Chow lemma is equivariant. Completeness again is only essential to avoid poles of V "hidden at ∞".

We may clearly assume X is normal, replacing X if necessary by its normalization and lifting V (cf. 1.9).

Hence we may imbed X equivariantly in $\mathbb{P}^N$ (1.8). Let W be the corresponding vector field on $\mathbb{P}^N$ and let $p \in \mathbb{P}^N$ be a zero of W. The projection from p $\mathbb{P}^N \longrightarrow \mathbb{P}^{N-1}$ is W equivariant. (Selecting coordinates so that $p = (0,\ldots,0,1)$ the projection is $(a_0,\ldots,a_N) \longrightarrow (a_0,\ldots,a_{N-1})$ but since p is a zero for $W = \sum_{ij} c_{ij} X_j \frac{\partial}{\partial X_i}$, i.e. $\sum_i c_{in} \frac{\partial}{\partial X_i} = \lambda \frac{\partial}{\partial X_n}$, we see that $c_{in} = 0$ for $i < n$ and hence W yields a well defined vector field W_1 on $\mathbb{P}^{N-1}$). Denote by X_1 the closure of the image of X in $\mathbb{P}^{N-1}$. W_1 is tangent to X_1. Note that either $\dim(X_1) = n$ or $\dim(X_1) = n-1$ and this second case can only arise if every line through p meeting X in a point $q \neq p$ lies entirely in X. If $\dim X_1 = n$, project equivariantly from p_1, $\mathbb{P}^{N-1} \longrightarrow \mathbb{P}^{N-2}$ obtaining X_2 and W_2 and continue in this manner until $\dim(X_r) = n-1$ (while $\dim(X_{r-1} = n)$. This will surely occur for $r \leq N-n+1$.

Note that the vector field W_r is substantial on X_r since the equivariant rational map $X \longrightarrow X_r$ has dense image (1.15). Moreover, W_r has zeroes on X_r, since it has zeroes on $\mathbb{P}^{N-r}$ (cf. 1.7). Hence by induction X_r is equivariantly birational to $\mathbb{P}^{n-1}$. It is now evident that X_{r-1} is birational to $\mathbb{P}^n$. Indeed, the fibres of the projection $X_{r-1} \longrightarrow X_r$ are simply the lines through p, whence X_{r-1} is exhibited as a $\mathbb{P}^1$ bundle over the rational variety X_r and is clearly rational. To be more precise, and to obtain equivariance, we describe X_{r-1} more precisely, namely if $\mathcal{O}(1)$ denotes the canonical line bundle on $X_r \subseteq \mathbb{P}^{N-r}$ then X_{r-1} is equivariantly birational to $\mathbb{P}(\mathcal{O} \oplus \mathcal{O}(1)) \longrightarrow X_r$, (i.e. if $\hat{\mathbb{P}}$ denotes $\mathbb{P}^{N-r+1}$ blown up at p_{r-1} and $\hat{X}_{r-1}$ denotes the proper transform of X_{r-1}, then $\hat{\mathbb{P}}$ is $\mathbb{P}(\mathcal{O} \oplus \mathcal{O}(1))$ over $\mathbb{P}^{N-r}$ and $\hat{X}_{r-1}$ is just the restriction of this bundle over X_r. This monoidal transform is equivariant (1.10)).

The proof that X_{r-1} is <u>equivariantly</u> birational to $\mathbb{P}^n$ is completed by the following two lemmata.

Lemma 3.1: Given $f : \mathbb{P}^{n-1} \longrightarrow Y$ equivariantly birational and given $Z = \mathbb{P}(\mathcal{O} \oplus L')$ for $L' \longrightarrow Y$ a line bundle then Z is equivariantly birational to $\mathbb{P}(\mathcal{O} \oplus L)$ for $L \longrightarrow \mathbb{P}^{n-1}$ a suitable line bundle.

Lemma 3.2: Given a substantial vector field V on $\mathbb{P}(\mathcal{O} \oplus L)$ for $L \longrightarrow \mathbb{P}^{n-1}$ a line bundle, there is an equivariant birational map $\mathbb{P}(\mathcal{O} \oplus L) \longrightarrow \mathbb{P}^n$.

Given the equivariant birational map $X_{r-1} \longrightarrow \mathbb{P}^n$ we turn to the study of the induced equivariant, rational, dominant map $X \longrightarrow \mathbb{P}^n$ in §4.

Proof of 3.1: We may assume Y is nonsingular (replacing Y by an equivariant resolution $\hat{Y} \longrightarrow Y$ and replacing Z by $Z \times_Y \hat{Y}$). Moreover, since Y is birational to $\mathbb{P}^{n-1}$ we know $H^1(Y, \mathcal{O}_Y) = 0$. Hence a vector field on Z is determined by its image V' on Y and by giving $D' : \mathcal{O} \oplus L' \longrightarrow \mathcal{O} \oplus L'$ satisfying $D'(g\sigma) = V'(g)\sigma + gD'(\sigma)$ for g a function and σ a section of $E' = \mathcal{O} \oplus L'$, (cf. 1.2, 1.3, noting that $H^1(\mathrm{Hom}(E',E')) \longrightarrow H^1(\mathrm{Hom}(E',E')|\mathcal{O})$ is injective).

Consider the rational map $f : \mathbb{P}^{n-1} \longrightarrow Y$ defined on an open set $U \subseteq \mathbb{P}^{n-1}$ whose complement has codimension ≥ 2. The line bundle $f^*(L')$ on U extends uniquely to a line bundle L on $\mathbb{P}^{n-1}$, ([7_2], Exp. XI, §3, pp. 126-130). Clearly, Z is birational to $\mathbb{P}(\mathcal{O} \oplus L)$ over $\mathbb{P}^{n-1}$. To check equivariance we must extend the derivation D' from $\mathcal{O} \oplus L|_U$ to $\mathcal{O} \oplus L$. Note that V' extends to a vector field V on $\mathbb{P}^{n-1}$ (since f is equivariant). Now given any point $p \in \mathbb{P}^{n-1} - U$ one may uniquely extend D' to a neighborhood W of p by selecting a basis e_1, e_2 for $\mathcal{O} \oplus L$ at p, then noting that $D'(e_i) = \Sigma a_{ij} e_j$ for a_{ij} being holomorphic functions on $(P^{n-1} - U) \cap W$ which extend uniquely to holomorphic functions on W (since $\mathrm{cod}(P^{n-1} - U) \geq 2$).

Proof of 3.2: Given a substantial vector field on $\mathbb{P}(\mathcal{O} \oplus L)$ over $\mathbb{P}^{n-1}$ we denote by V the induced vector field on $\mathbb{P}^{n-1}$. Note that V is substantial (1.15). We chose homogeneous coordinates X_i so that either (cf. 2.1)

Case 1: $V = X_0 \, \partial/\partial X_1 + \sum_{i>1} \lambda_i X_i \, \partial/\partial X_i$ or

Case 2: $V = \sum_{i>0} \lambda_i X_i \, \partial/\partial X_i$

The vector field on $\mathbb{P}(\mathcal{O} \oplus L)$ is given by a V-derivation $D : \mathcal{O} \oplus L^* \longrightarrow \mathcal{O} \oplus L^*$. We may assume $L = \mathcal{O}(k)$ for $k \geq 0$ since $\mathbb{P}(\mathcal{O} \oplus L) \xrightarrow{\sim} \mathbb{P}(\mathcal{O} \oplus L^*)$.

In terms of a local basis e for $L^* = \mathcal{O}(k)$ one may determine D locally by $D(1) = a \cdot 1 + b \cdot e$; $D(e) = c \cdot 1 + d \cdot e$ and changing the choice of e one finds that the b's (resp. c's) define a global section of $\mathcal{O}(k)$ (resp. $\mathcal{O}(-k)$), and "a" is a scalar. Thus $c = 0$ if $k > 0$ and L^* is necessarily a D invariant sub-bundle. If $k = 0$, then choose a suitable basis for $\mathcal{O} \oplus L^* = \mathcal{O} \oplus \mathcal{O}$ so that D has upper triangular form, i.e. again $c = 0$ and L^* is invariant. Thus, in either case D defines a V-derivation of L^*. The section X_0^k of L^* is an eigen-section for D, in view of the form of V. Normalizing D (subtracting off a scalar multiple of the identity) one may assume $D(X_0^k) = 0$.

Thus on the affine $X_0 \neq 0$, one may take $e = X_0^k$, so that $D(e) = 0$ and $D(1) = a\cdot 1 + b\cdot e$, or letting $t = 1/e$ be the fibre coordinate

$$D(t) + at + b$$

where "a" is a scalar and $b = b(x)$ a polynomial of degree $\leq k$ in the affine coordinates $x_i = X_i X_0^{-1}$.

We show below that for a suitable choice of fibre coordinate t' the rational mapping $\mathbb{P}(\mathcal{O} \oplus \mathcal{O}(L) \longrightarrow \mathbb{P}^n$ defined by $(1, x_1, \ldots, x_{n-1}, t')$ is equivariant (carrying our vector field to a global holomorphic field on $\mathbb{P}^n$). We let Y_i denote homogeneous coordinates on $\mathbb{P}^n$, and $y_i = Y_i/Y_0$.

Case I: $V = \partial/\partial x_1 + \sum_{i>1} \lambda_i x_i \partial/\partial x_i$; $D(t) = at + b$

We seek a polynomial $q(x)$ such that if $t' = t - q$ then $D(t') = a t'$. Given t', V would correspond to the vector field

$$\partial/\partial y_1 + \sum_{i=1}^{n-1} \lambda_i y_i \partial/\partial y_i + a y_n \partial/\partial y_n \quad \text{on} \quad P^n$$

Such a $q(x)$ is precisely a solution to the equation $V(q) - a q = b$. Expanding $q = \Sigma q_I x^I$, $b = \Sigma b_I x^I$ one obtains the equations

$$(i_1 + 1) q_{I+e_1} + (\lambda\cdot\tilde{I})\, q_I + a\, q_I = b_I \quad \text{for each}$$

$I = (i_1, \ldots, i_{n-1})$ where $\tilde{I} = (i_2, \ldots, i_{n-1})$ and $(I + e_1) = (i_1 + 1, i_2, \ldots, i_{n-1})$. Note that all multiindices appearing in this equation have the same $\tilde{I}$. Hence for each fixed $\tilde{I}$ one has an independent system of equations. We fix $\tilde{I}$ and solve. If $(\lambda\cdot\tilde{I} + a) = 0$ then the equations become $(i_1 + 1)\, q_{I+e_1} = b_I$ and are readily solved. If $(\lambda\cdot\tilde{I} + a) \neq 0$, then recursively solve for the q_I descending on i_1 and defining $q_I = 0$ if $b_{I'} = 0$ for all $i_1' \geq i_1$, $\tilde{I} = \tilde{I}'$. Then the q so obtained is clearly polynomial.

Case II: $V = \sum_{i>0} \lambda_i x_i \partial/\partial x_i$, $D(t) = at + b$.

Again we seek $q(x)$ such that $D(t - q) = a(t - q)$. We obtain equations

$$(I\cdot\lambda - a) q_I = b_I .$$

For those I such that $I\cdot\lambda \neq a$ we solve the equations for q_I defining $q_I = 0$ otherwise and set $t' = t - q$. Then $D(t') = at' + \tilde{b}$ where $\tilde{b} = \Sigma \tilde{b}_I x^I$ with $\tilde{b}_I = 0$ unless $I\cdot\lambda = a$, i.e. $V(\tilde{b}) = a\tilde{b}$. If $\tilde{b} = 0$ we have succeeded, if not we redefine t' as $t' = (t - q)/\tilde{b}$. Then $D(t') = 1$. The birational map defined by x_i and this t' transports our vector field to $\sum_{i=1}^{n-1} \lambda_i y_i \partial/\partial y_i + \partial/\partial y_n$ which is globally holomorphic on $\mathbb{P}^n$.

§4. Analysis of the map $X \longrightarrow \mathbb{P}^n$.

Consider the rational dominant equivariant map $X \longrightarrow \mathbb{P}^n$ constructed on §3. One may obtain $\hat{X} \longrightarrow X$ by a sequence of nomoidal transformations with non-singular invariant centers so that $\hat{X}$ is nonsingular and the induced rational map $f: \hat{X} \longrightarrow \mathbb{P}^n$ extends to a globally defined holomorphic map (by employing Hironaka's equivariant resolution). Consider $Y \subseteq P^n$ the analytic subset which is the image of the locus where df is not of maximal rank. Clearly Y is invariant, and $\hat{X} - f^{-1}(Y) \longrightarrow P^n - Y$ is a connected covering with finitely many sheets. Note that for $d >> 0$ there exist hypersurfaces of degree d in $\mathbb{P}^n$ which contain Y.

One may select such a hypersurface H which is invariant. Indeed, the set of hypersurfaces of degree d which contain Y is an invariant nonempty subset of the projective space of degree d hypersurfaces. By 1.7 (Borel-Sommese) there exists an invariant point H in this set. Again $\hat{X} - f^{-1}(H) \longrightarrow \mathbb{P}^n - H$ is a connected finitely sheeted covering map.

However, since $X \longrightarrow \mathbb{P}^n$ is equivariant and dominant the induced vector field on $\mathbb{P}^n$ is substantial. The homogeneous defining equation for H is an eigenfunction for this vector field and is hence a monomial (cf. 2.2). Thus H is a union of coordinate hyperplanes and $\mathbb{P}^n - H \xrightarrow{\sim} \mathbb{C}^{n-r} \times (\mathbb{C}^*)^r$, where H is set theoretically the union of $r+1$ hyperplanes. The fundamental group of $\mathbb{P}^n - H$ is then $\mathbb{Z}^{(r)}$ and the algebraic isomorphism classes of connected algebraic covers of $\mathbb{P}^n - H$ are in 1-1 correspondence with subgroups of finite index in $\mathbb{Z}^{(r)}$ ([7], expose 12, 5.1).

Such a subgroup is specified by giving a nonsingular $r \times r$ matrix of integers $((n_{ij}))$ (the columns of this matrix being a set of basic generators for the subgroup) with two matrices corresponding to the same subgroup if and only if they differ by post multiplication by $GL(r,\mathbb{Z})$. Given $((n_{ij}))$ the corresponding cover is

4.1: $$\mathbb{C}^{n-2} \times \mathbb{C}^{*r} \longrightarrow \mathbb{C}^{n-r} \times \mathbb{C}^{*r}$$
$$(x,y_1,\ldots,y_r) \longrightarrow (x, y_1^{n_{11}} y_2^{n_{12}} \cdots y_r^{n_{1r}}, \ldots, y_1^{n_{r1}} y_2^{n_{r2}} \cdots y_r^{n_{rr}}).$$

Hence $\hat{X} - f^{-1}(H)$ is algebraically isomorphic to $\mathbb{C}^{n-r} \times \mathbb{C}^{*r}$ and therefore $\hat{X}$ is rational, as is X. To complete the inductive step we must verify that the rational map $\hat{X} \longrightarrow \mathbb{P}^n$ envisaged, i.e. $(1, x_1, \ldots, x_{n-r}, y_1, \ldots, y_r)$ is <u>equivariant</u>.

We explicitly compute (in terms of the affine coordinates on $\hat{X}$ denoted by x,y as in 4.1) the vector field $\hat{V}$ lifting V on $\mathbb{P}^n$ (which will be expressed in affine coordinates u,v, according to the $\mathbb{C}^{n-r}$, $\mathbb{C}^r$ of 4.1). We must verify that $\hat{V}$ closes up to a holomorphic vector field on $\mathbb{P}^n$.

Recall that since V is substantial, it had the form $\Sigma \lambda_i X_i \partial/\partial X_i$ or $X_0 \partial/\partial X_1 + \sum_{i>1} \lambda_i X_i \partial/\partial X_i$. In the second case the monomial defining H will not involve the variable X_1, (cf. §2) so that X_1/X_0 is one of the "u" variables.

To lift V , note that the vector fields $u_i\ \partial/\partial u_i$ and $\partial/\partial u_i$ lift to $x_i\ \partial/\partial x_i$ and $\partial/\partial x_i$. Moreover, the image of $y_j\ \partial/\partial y_j$ is $\sum_i n_{ij}\ v_i\ \partial/\partial v_i$. Hence the vector field $\Sigma\ \lambda_i\ v_i\ \partial/\partial v_i$ is the image of $\Sigma\ \eta_j\ y_j\ \partial/\partial y_j$ where $\eta = N^{-1}\lambda$, $N = ((\eta_{ij}))$.

The lifted vector field clearly extends to a holomorphic vector field on $\mathbb{P}^n$, exhibiting the equivariant birationality of $\hat{X}$ and $\mathbb{P}^n$, and completing the inductive proof.

Remark: The more unpleasant aspects of the proof of §3 and §4 lay in showing that given a certain X birational to $\mathbb{P}^n$ and a global holomorphic vector field on X , then X was equivariantly birational to $\mathbb{P}^n$ (not perhaps by the given rational map). A simple example where the given rational map is inadequate but a different birational map is equivariant may be obtained by considering a quadric Q in $\mathbb{P}^{N+1}$ Projection from a point q on Q yields a birational map $Q \longrightarrow \mathbb{P}^n$. Fixing any vector field V on $\mathbb{P}^{n+1}$ which is tangent to Q the birational map is equivariant if and only if q is a zero of V . The question of whether there always exists an equivariant birationality is an interesting one.

BIBLIOGRAPHY

[1] Atiyah, M., Complex analytic connections in fibre bundles, Trans. Am. Math. Soc. 85, (1957) 181-207.

[2] Blanchard, A., Un théorème sur les automorphismes d'un variété complexe projective, C.R. Acad. Sci. 240 (1955) 2198-2201.

[3] Burns, D. and Wahl, J., Local contributions to global deformations of surfaces, Invent. Math. 26 (1974) 67-88.

[4] Carrell, J., Howard, A., Kosniowski, C., Holomorphic vector fields on complex surfaces, Math. Ann. 204 (1973) 73-82.

[5] Carrell, J., Lieberman, D., Holomorphic vector fields and Kaehler manifolds, Invent. Math. 21 (1973) 303-309.

[6] Grothendieck, A., Eléments de géométrie algébrique, I, ch.0, IHES 4, 1960.

[7] ______________, Séminaire de géométrie algébrique I, (1960-61).

[7_2] ______________, Cohomologie locale des faisceaux cohérents et théorèmes de Lefschetz locaux et globaux (SGA 2), Advanced Studied in Math., V.2, North Holland, 1968.

[8] Hironaka, H., Desingularization of complex analytic varieties, Actes du Congr. Internat. Math. vol. 2, 1970, 627-633.

[9] Howard, A., Holomorphic vector fields on algebraic manifolds, Amer. J. Math., 94 (1972), 1282-1290.

[10] Kobayashi, S., Transformation groups in differential geometry, New York: Springer, 1972.

[11] Lichnerowcz, A., Variétés Kahlériennes et première class de Chern, J. of Diff. Geom. 1, (1967) 195-224.

[12] Matsushima, Y., Holomorphic vector fields and the first Chern class of a Hodge manifold, J. of Diff. Geom. 3, (1969) 477-480.

[13] Poincaré, H., Oeuvres, vol. 1, XCIX-CV.

[14] Seidenberg, A., Derivations and integral closure, Pac. J. Math. 16 (1966), 167-174.

[15] Shafarevic, I.R., et al., Algebraic Surfaces, Proc. Steklov, Inst. Math., 1965, or Amer. Math. Soc., 1967.

[16] Sommese, A., Holomorphic vector fields on compact Kaehler manifolds, Math. Ann. 210 (1974) 75-82.

SOME EXAMPLES OF $\mathbb{C}^*$ ACTIONS

by

Andrew John Sommese

In this note I give examples of $\mathbb{C}^*$ actions whose Bialynicki-Birula decompositions exhibit a number of bad properties.

Let $r : \mathbb{C}^* \times X \to X$ denote a meromorphic action of $\mathbb{C}^*$ on a compact connected complex manifold X; i.e. r is a holomorphic action that extends to a meromorphic map $\mathbb{P}^1_{\mathbb{C}} \times X \to X$. Let $F_1, \ldots, F_r$ be the connected components of $X(\mathbb{C}^*)$, the fixed point set of the action. Under the above hypotheses the maps:

$$A^+ : X \to X(\mathbb{C}^*) \qquad A^- : X \to X(\mathbb{C}^*)$$

defined by:

$$A^+(x) = \lim_{t\to 0} r(t,x) \quad \text{and} \quad A^-(x) = \lim_{t\to\infty} r(t,x)$$

are well defined. Let:

$$X_i^+ = A^{+\,-1}(F_i) \quad \text{and} \quad X_i^- = A^{-\,-1}(F_i) .$$

These sets are constructible. They are called respectively, the plus and minus components over X_i. There are two distinct fixed point components denoted F_1 and F_r called respectively the source of the action and the sink of the action that are characterized by the properties that X_1^+ and X_r^- are dense Zariski open sets of X.

The Bialynicki-Birula plus and minus decompositions [1, 2, 5, 8, 11, 12] are:

$$X = \bigcup_i X_i^+ \qquad \text{and} \qquad X = \bigcup_i X_i^-$$

respectively. These above functorial decompositions that generalize the classical Bruhat decomposition play a key role [2, 3, 4, 6, 7, 8] in the study of $\mathbb{C}^*$ actions. If X is either Kaehler or algebraic, the above decompositions enjoy a number of good properties. Two of the most important properties are:

a) the maps $A^+ : X_i^+ \to F_i$ and $A^- : X_i^- \to F_i$ are continuous (and in fact) holomorphic maps,

and,

b) the sets X_i^+ and X_i^- are locally closed.

In §1 I give an example (cf. (1.3)) of a <u>meromorhpic $\mathbb{C}^*$ action on a compact complex manifold, X, that is bimeromorphic to $\mathbb{P}^3_{\mathbb{C}}$ and for which the maps in a) are discontinuous and b) fails</u>. It is also shown (cf. (1.1)) that b) can hold while the maps in a) are discontinuous. The examples are variations on Hironaka's famous example [9, pg. 441ff] of an algebraic non-projective manifold.

By the same method examples are given in §1 that exhibit the phenomenon first discovered by Jurkiewicz [10] of non-trivial cycles of orbits. Recall that Jurkiewicz produced an example of an algebraic $\mathbb{C}^*$ action on an algebraic manifold X with a sequence of non-fixed points $x_1, x_2, \ldots, x_n$ with the property that:

#) $A^+(x_i) = A^-(x_{i-1})$ for $i = 1$ to n (with the convention that $x_0 = x_n$).

In Jurkiewicz's example constructed by torus embeddings $n \geq 7$. For algebraic manifolds Bialynicki-Birula pointed out that n must be at least 2. I give a very simple example (1.1) of an algebraic $\mathbb{C}^*$ action on an algebraic manifold with the property #) and $n = 2$, the minimum possible. I also give an example (1.2) of a meromorphic action of $\mathbb{C}^*$ on a Moisezon manifold, X, for which $n = 1$ occurs, i.e. there is a point $x \in X$ which is not in $X(\mathbb{C}^*)$ and such that $A^+(x)$ and $A^-(x)$ both belong to the same fixed point component.

In §1 I also make a conjecture which if true will put a definite limit on the pathologies that can be expected for meromorphic $\mathbb{C}^*$ actions on compact complex manifolds.

In §2 I give an example that is surprising in light of the results of [3].

Most of these examples were worked out during the stimulating conference organized by James Carrell with NSERC funds at the University of British Columbia during the period January 15, 1981 to February 15, 1981. I would like to thank the National Science Foundation, the University of Notre Dame, and the Sloan Foundation for their support during parts of the period when this research was carried out. These examples would not have been worked out without the encouragement of Andrzej Bialynicki-Birula. I would also like to thank David Liebermann for providing some extra impetus to construct the examples in §1.

§1 The Main Examples

I start by giving Hironaka's example [9, pg. 441ff] with attention to the fact that it can be presented so that it comes equipped with a $\mathbb{C}^*$ action.

(1.1) Example. Let S denote $\mathbb{P}^2_{\mathbb{C}}$ and let $M = S \times \mathbb{P}^1_{\mathbb{C}}$. Let r' denote the action on M given by:

$$r'(t, s, [z_0, z_1]) = (s, [z_0, tz_1])$$

where $[z_0, z_1]$ denotes homogeneous coordinates. Let 0 denote $[1,0]$ and let S_0 denote $S \times 0$, the source of the action r'. Let C denote a line on S and let D denote a smooth conic on S that meets C in two distinct points, x, y. Cover S with three open sets U, V, and W such that x doesn't belong to $\overline{V} \cup \overline{W}$ and y doesn't belong to $\overline{U} \cup \overline{W}$. Let U', V', and W' denote the inverse images of U, V, and W respectively under the projection $M \to S$. Under the projection from M to S, I identify C, D, x, and y with the corresponding curves and points on S_0. Blow up $(C \cup D) \cap W'$ to get a complex manifold W''. Blow up $C \cap U'$ to get a complex manifold U''. Blow up the proper transform of $D \cap U'$ in

U" to get a complex manifold $\mathcal{U}$. Blow up $D \cap V'$ to get a complex manifold V". Blow up the proper transform of $C \cap V'$ to get a complex manifold $\mathcal{V}$. It is easy to check that the three complex manifolds $\mathcal{U}$, $\mathcal{V}$, and W" patch to give a compact complex manifold X. This manifold is algebraic and birational to M and hence $\mathbb{P}^3_{\mathbb{C}}$ [9, pg. 441ff]. Using the following standard lemma repeatedly the reader can check that the action r' lifts to a meromorphic action of $\mathbb{C}^*$ on X.

(1.1.1) Lemma. Let $r : \mathbb{C}^* \times Y \to Y$ denote a holomorphic action of $\mathbb{C}^*$ on a complex manifold Y. Let Y' denote Y with a complex submanifold B blown up. If $r(\mathbb{C}^*, B) = B$ then the action r lifts to a holomorphic action r' on Y'. If r is a meromorhpic action then so is r'.

The above example has 4 fixed point components. Besides the source and the sink there are two components biholomorphic to $\mathbb{P}^1_{\mathbb{C}}$. Under the induced projection from $p : X \to S$ these components F and G go biholomorphically onto C and D respectively. There is a point $x' \in p^{-1}(x)$ such that $A^+(x') \in F$ and $A^-(x') \in G$. There is a point $y' \in p^{-1}(y)$ such that $A^+(y') \in G$ and $A^-(y') \in F$.

It is easy to modify the above example to get X, a Moisezon manifold birational to $\mathbb{P}^3_{\mathbb{C}}$ with a meromorphic $\mathbb{C}^*$ action possessing a point $x' \in X - X(\mathbb{C}^*)$ such that $A^+(x')$ and $A^-(x')$ belong to the same fixed point component. Note that in such an example the property a) of the Bialynicki-Birula decomposition that was discussed in the introduction fails. The discontinuity occurs because by continuity $A^-(x')$ should go to $A^+(x')$ and not to $A^+(A^-(x')) = A^-(x')$. Let me sketch this construction.

(1.2) Example. Let S, M, S_0 and r' be as in the last example. Let C be an irreducible cubic curve on S which has precisely one singularity x. Assume that x is a node, i.e. C is gotten from $\mathbb{P}^1_{\mathbb{C}}$ by identifying two points. Cover S with two open sets U and V such that x doesn't belong to $\overline{V}$ and such that $C \cap U$ is the union of two irreducible components C_1 and C_2 which are both smooth Let U' and V' be as in the last example. Identify C and x with the corresponding curve and point on S_0. Blow up $C \cap V'$ to get a complex manifold V". Blow up C_1 to get a complex manifold U". Blow up the proper transform of C_2 in U" to get a complex manifold $\mathcal{U}$. It is easily checked that V" and $\mathcal{U}$ patch to give the desired example.

The above construction can be modified further to yield an example where both of the properties a) and b) of the Bialynicki-Birula decomposition that were discussed in the introduction fail.

(1.3) Example. Let C and D be two irreducible curves on a rational surface S. Assume that:

1) C and D meet only in a point x,

2) D is smooth and x is the only singular point of C,

3) there is an open set U containing x such that $C \cap U$ consists of two

components C_1 and C_2 that are both smooth and meet transversely at x,

4) D meets both C_1 and C_2 transversely at x.

It is easy to find such a triple (S, C, D). Indeed let D' be a smooth line on $\mathbb{P}^2_{\mathbb{C}}$ and let C' be a cubic curve on $\mathbb{P}^2_{\mathbb{C}}$ as in the last example. Assume that C' and D' meet at the singular point of C' in a manner satisfying properties 2), 3) and 4) above. C' and D' also meet in a point y besides the singular point of C'. Let S be $\mathbb{P}^2_{\mathbb{C}}$ with y blown up. Let C and D be the proper transforms of C' and D' respectively.

Let $M = S \times \mathbb{P}^1_{\mathbb{C}}$ and let $\mathbb{C}^*$ act on M as in the past examples. Let V be an open set on S such that $U \cup V = S$ and $x \notin \overline{V}$. Let U' and V' be the open sets on M as in the preceding examples. Identify C, D, and x with curves and a point of the same name on $S_0 = S \times 0$. Blow up $(C \cup D) \cap V'$ to get a complex manifold V''. Blow up $C_1 \cap U'$ to get a complex manifold U''. Blow up the proper transform of $D \cap U'$ on U'' to get a complex manifold $U^{\#}$. Blow up the proper transform of $C_2 \cap U'$ on $U^{\#}$ to get a complex manifold $\mathcal{U}$. $\mathcal{U}$ and V'' patch to give the desired Moisezon manifold X. By lemma (1.1.1) the $\mathbb{C}^*$ action on M lifts to a meromorphic action:

$$r : \mathbb{C}^* \times X \longrightarrow X .$$

There are two fixed point components F and G for this action that map birationally onto C and D respectively under the projection

$$f : X \longrightarrow S$$

induced from $M = S \times \mathbb{P}^1_{\mathbb{C}} \to S$. Let Z_x denote the fibre of $X \to S$ over x under the map f. There are points:

$$\{x_1, x_2, x_3\} \subseteq Z_x - X(\mathbb{C}^*)$$

such that $A^+(x_3) = A^-(x_2) = z \in G$, $A^-(x_3) \in F$, and $A^+(x_2) = A^-(x_1) \in F$. It is a straightforward topological check that the minus component F^- over F is not locally closed at z. Nor is the restriction of A^- to F^- continuous at z. The following picture might help:

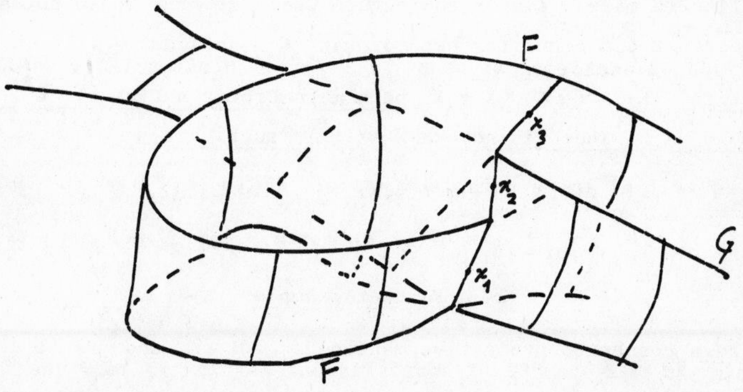

(1.4) To end this section I would like to point out a pathology that follows by carrying the above constructions a little further. Let $r' : \mathbb{C}^* \times X \to X$ be a meromorphic $\mathbb{C}^*$ action on a compact complex manifold. By a principle orbit is meant an orbit $0 = r(\mathbb{C}^*, x)$ such that $A^+(0)$ belongs to the source and $A^-(0)$ belongs to the sink of the action. It makes sense [cf. 8, 3] to consider limits of principle orbits. Such a limit consists of a sequence of orbits $\{0_1, \ldots, 0_k\}$ such that $A^+(0_j) = A^-(0_{j-1})$ for $j = 2, \ldots, k$. For example Z_x in the last example minus its intersections with the source and with the sink is a limit of principle orbits. Fix a finite set of letters a, b, ... and any finite word in them, e.g. for concreteness abracadabra. Then an example can be constructed of a Moisezon manifold birational to $\mathbb{P}^3_{\mathbb{C}}$ with a meromorphic $\mathbb{C}^*$ action $r : \mathbb{C}^* \times X \to X$ such that:

a) the fixed point components of r are the source S_0, the sink S_∞, and components indexed by a , b , ...,

b) there is a limit of principle orbits $\{0_1, \ldots, 0_k\}$ with the subscripts of the fixed point sets to which $A^-(0_j)$ belong for $j = 1, \ldots, k-1$, spelling the given word.

To construct this example choose a rational surface S with a finite set $H_1, \ldots, H_r$ of irreducible curves, and an assignment of one letter to each of the H_i. The required properties of these curves are:

1) for different i and j, H_i and H_j meet only in a point x that is independent of i and j,

2) there is an open set $U \subseteq S$ such that for each i, $H_i \cap U$ is the union of $d(i)$ smooth curves, $H_{i,d(1)}, \ldots, H_{i,d(i)}$, each two of which meet transversely at x,

3) for different i and j, each component of $H_i \cap U$ meets each component of $H_j \cap U$ transversely in x,

4) $d(i)$ equals the number of times the letter associated to H_i occurs in the word fixed at the beginning of the construction.

A surface S and curves satisfying the above properties can always be constructed by blowing up and blowing down. By blowing up the $H_{i,j}$ in the order prescribed by spelling the specified word backwards, the example is constructed.

I would like to make a simple conjecture which if true would put a limit on the pathologies that can occur for meromorphic $\mathbb{C}^*$ actions.

(1.5) <u>Conjecture</u>. <u>Let</u> $r : \mathbb{C}^* \times X \to X$ <u>be a meromorphic action of</u> $\mathbb{C}^*$ <u>on a compact complex manifold</u> X. <u>There is no</u> $x \in X - X(\mathbb{C}^*)$ <u>such that</u>:

$$\lim_{t \to 0} r(t,x) = \lim_{t \to \infty} r(t,x) .$$

§2 A Further Example

Let Y be a connected projective manifold. Let A and B be two projective submanifolds of Y. Let Z denote the set theoretic intersection of A and B. Let $M = \mathbb{P}^1_{\mathbb{C}} \times X$. Let $A_0 = [1,0] \times A$ and let B_∞ denote $[0,1] \times B$. Blow up A_0 and B_∞ to get a complex manifold X. The meromorphic $\mathbb{C}^*$ action on M along the $\mathbb{P}^1_{\mathbb{C}}$ lifts to a meromorphic action r of $\mathbb{C}^*$ on X. There are two fixed point components F and G for this action besides the source and the sink. F is biholomorphic to A and G is biholomorphic to B. Let F^+ denote the plus component over F and let G^- denote the minus component over G. It is a straightforward check that the set theoretic intersection $F^+ \cap G^-$ is biholomorphic to $\mathbb{C}^* \times Z$.

The above shows that the sets $X_i^+ \cap X_j^-$ for a $\mathbb{C}^*$ action can be rather unpleasant, e.g. disconnected and singular, even when X is a projective manifold. This is not so surprising by itself but points out the unexpected simplicity of the results of [3]. There it is shown that $\mathbb{C}^*$ invariant open sets $X - X(\mathbb{C}^*)$ with compact complex geometric quotients are built out of the sets $X_i^+ \cap X_j^-$ (and not their connected components!).

List of References

[1] Bialynicki-Birula, A., Some theorems on actions of algebraic groups, Ann. of Math. 98 (1973), 480-497.

[2] Bialynicki-Birula, A., On fixed points of torus actions on projective varieties, Bull. de l'Acad. Pol. des Sci. ser. des sciences math., astr., et phys. XXII (1974), 1097-1101.

[3] Bialynicki-Birula, A. and Sommese, A.J., Quotients by torus actions, to appear.

[4] Bialynicki-Birula, A. and Sommese, A.J., On quotients by SL(2) actions, to appear.

[5] Carrell, J.B. and Sommese, A.J., $\mathbb{C}^*$ Actions, Math. Scand. 43 (1978), 49-59.

[6] Carrell, J.B. and Sommese, A.J., Some topological aspects of $\mathbb{C}^*$ actions on compact Kaehler manifolds, Comment. Math. Helv. 54 (1979), 567-582.

[7] Carrell, J.B. and Sommese, A.J., SL(2,$\mathbb{C}$) actions on compact Kaehler manifolds, to appear Trans. Amer. Math. Soc.

[8] Fujiki, A., Fixed points of torus actions on compact Kaehler manifolds, Publ. RIMS, Kyoto Univ. 15 (1979), 797-826.

[9] Hartshorne, R., Algebraic Geometry, Springer Verlag, New York (1977).

[10] Jurkiewicz, J., An example of algebraic torus action which determines a non-filterable decomposition, Bull. de l'Acad. Pol. des Sci. ser. des sciences math., astr., et phys. XXIV (1976) 667-674.

[11] Konarski, J., Decompositions of normal algebraic variaeties determined by an action of a one dimensional torus, Bull. de l'Acad. Pol. des Sci., ser. des sciences math. astr. et phys. XXVI (1978), 295-300.

[12] Koras, M., Actions of reductive groups, University of Warsaw doctoral thesis (1980).

Department of Mathematics
University of Notre Dame
Notre Dame, Indiana 46656
U.S.A.

THE GROWTH FUNCTION OF A DISCRETE GROUP

Philip Wagreich*

Suppose G is a finitely generated group and S is a set of generators. Assume for simplicity, that if $s \in S$ then $s^{-1} \in S$. One can define a function

$$\ell_S : G \dashrightarrow Z$$

by

$$\ell_S(g) = \inf\{n \mid g = s_1 \cdots s_n\,,\ s_i \in S\}$$

i.e., $\ell_S(g)$ is the length of the shortest word in the generators which represents g. We call $\ell_S(g)$ the <u>length</u> of g (relative to S and note that $\ell_S(1) = 0$).

<u>Definition</u> (1.1): The <u>growth function</u> (or growth sequence) of G relative to S is defined to be

$$a_n = \mathrm{card}\{g \in G \mid \ell_S(g) = n\}$$

the number of elements of G of length n. An important tool for studying a sequence is the generating function for the series, hence we define

<u>Definition</u> (1.2): If G, S are as above then we define a formal power series

$$p_S(t) = \sum_{n=0}^{\infty} a_n t^n = \sum_{g \in G} t^{\ell(g)}$$

We call p_S the growth power series of G, S. If G is a finite group then this power series is a polynomial. We shall see that for many infinite groups it is a rational function.

The growth function was studied by Milnor in [Mi 1], where he showed that there is a relation between the curvature of a compact manifold and the growth of its fundamental group. The growth power series arises in many contexts, for example if G is Coxeter group and S is a natural set of generating reflections. In the second section of this paper we will give a survey of some of the surprising and beautiful results that have been proven about this function in certain special cases.

My interest in the growth function was inspired by the preprint of Alan Durfee's

---

*Research partially supported by grants from the National Science Foundation and the University of Illinois at Chicago Circle Research Board.

paper, '14 Characterizations of Rational Double Points and Simple Critical Points'. It was known that the simple singularities are the only singularities with finite monodromy group. Durfee conjectured that only the simple elliptic singularities have monodromy groups with polynomial growth (i.e., a_n bounded above by a polynomial function of n, see 2.9) and all others have exponential growth (see 2.9). To prove this conjecture I was led to study the growth functions of the simplest infinite Coxeter groups, namely hyperbolic triangle groups. These groups turned out to have such marvelous and surprising properties that I never got to study Durfee's conjecture. (It was proven by Looijenga shortly after Durfee's preprint appeared. The paper appeared as '15 Characterizations of Rational Double Points and Simple Critical Points'.)

We adopt the convention that Σ stands for the sum from $n = 0$ to ∞.

(1.3) The hunt for invariants.

One reason for studying growth functions is to find invariants of the group. There are not many strong results in this direction but there are some tantalizing hints. Serre [Se] has shown that for a Coxeter group G (see 1.6) and the standard Coxeter generating set S, the value of $p_S(t)$ at $t = 1$ is $1/\chi$ where χ is the Euler characteristic of the group G (this makes sense when p_S has a pole at t if we let $1/0 = \infty$). This result is trivially true for finite groups because $p_S(1)$ = order G and $\chi = 1/\text{order } G$. In all examples of infinite groups that I know the same equality holds.

Bass [Ba] has shown that if G is nilpotent then for any generating set $\{a_n\}$ has polynomial growth and the 'degree of polynomial growth' can be defined and is independent of S.

Finally, if G is a discrete transformation group and S is a set of generators having some geometric significance then the radius of convergence R of $p_S(t)$ should also have some geometric significance. If G has exponential growth then R determines the 'rate' of growth of $\{a_n\}$.

(1.4) <u>Triangle groups</u>.

Suppose p, q, r are positive integers and $1/p + 1/q + 1/r < 1$. Let T be a triangle in the hyperbolic plane with angles π/p, π/q and π/r. Let $\Delta_{p,q,r}$ be the group generated by the three reflections s_1, s_2, s_3 in the sides of T. Then $\Delta_{p,q,r}$ is a discrete discontinuous group of isometries of the hyperbolic plane

<u>Proposition</u> (1.5). If $S = \{s_1, s_2, s_3\}$ then $p_S(t)$ is a rational function, in fact

$$p_S(t) = \frac{(1+t)(1-t^p)(1-t^q)(1-t^r)}{1 - 2t + t^{p+1} + t^{q+1} + t^{r+1} - t^{p+q} - t^{p+r} - t^{q+r} + 2t^{p+q+r} - t^{p+q+r+1}}$$

This can be easily proven by noting that $\Delta_{p,q,r}$ is a Coxeter group and using the algorithm in [B] (see 2.8). There is an alternate proof [W] using facts about tesselations of the hyperbolic plane. This proof generalizes to Fuchsian groups. It will be discussed in §3. Note that the numerator and denominator of p_S are reciprocal polynomials i.e., if λ is a root then $1/\lambda$ is a root of the same multiplicity.

The location of the poles of p_S is of some interest. For example, G has exponential growth if and only if there is a pole inside the unit circle. It is also interesting to note that the coefficients of the numerator and denominator above are anti-palindromic. Theorem (1.10) below was conjectured after examining three pounds of computer printout from a program written by A.O.L. Atkin. (His program efficiently factors polynomials over $\mathbb{Z}$.) In order to state the theorem we must first introduce some notation.

(1.6) Coxeter groups (see [B]).

Suppose $n \in \mathbb{N}$. A Coxeter graph Γ is a graph with vertices $v_1, \ldots, v_n$ and for each pair of vertices at most one edge and an integer (or ∞) weight $m_{ij} > 3$ for each edge. If there is no edge from v_i to v_j we define $m_{ij} = 2$. We define $m_{ii} = 1$, for $i = 1, \ldots, n$. The matrix $M = (m_{ij})$ is called the Coxeter matrix associated to Γ. Note that M is symmetric. The Coxeter group G_Γ associated to Γ is the group with the following presentation:

generators: $s_1, \ldots, s_n$

relations: $(s_i s_j)^{m_{i,j}} = 1$, for all i, j.

Note that for $i = j$ we get $s_i^2 = 1$, so that G_Γ is generated by elements of order 2. Let $S_\Gamma = \{s_1, \ldots, s_n\}$.

Example (1.7). If $\Gamma =$

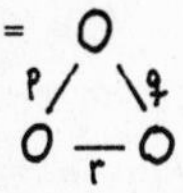

and $1/p + 1/q + 1/r < 1$,

then G_Γ is isomorphic to $\Delta_{p,q,r}$. If $1/p + 1/q + 1/r = 1$ (resp. > 1) then G_Γ is the group generated by the reflections in the sides of the euclidean (resp. spherical) triangle with angles π/p, π/q, π/r.

Example (1.8). Let $G_{a,b,c}$ be the Coxeter group associated to the graph

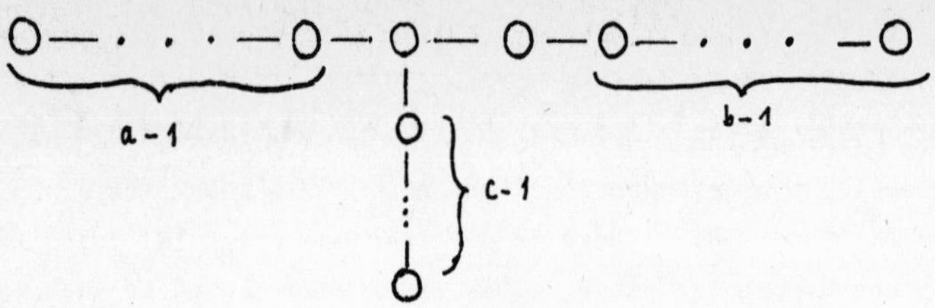

(all edges have weight 3). This group is finite if and only if $1/a + 1/b + 1/c < 1$. See [B] where it is shown that G_Γ is finite if and only if Γ is on a certain well known list of graphs, namely:

$$A_k, B_k, D_k, E_6, E_7, E_8, F_4, G_2, H_3, H_4, I_2(p)$$

<u>Definition</u> (1.9). A Coxeter element of a Coxeter group G is an element of the form $g = s_1 s_2 \ldots s_n$, where the s_i are the generators defined above.

Any two Coxeter elements are conjugate, thus if G is finite h = order G is a well defined integer called the Coxeter number of G.

<u>Theorem</u> (1.10): [W] Let $G = \Delta_{p,q,r}$ with $1/p + 1/q + 1/r < 1$ and $S = \{s_1, s_2. s_3\}$ as above. Let $p_S(t) = f(t)/g(t)$ with f and g relatively prime. Then

1. all but one of the irreducible factors of g is cyclotomic.
2. suppose ξ is a primitive n^{th} root of unity. Then $g(\xi) = 0$ if and only if the following three conditions hold:
   a) n does not divide p, q, and r.
   b) $1/\overline{p} + 1/\overline{q} + 1/\overline{r} > 1$ (where $\overline{x}$ denotes the remainder of x after division by n).
   c) n is the Coxeter number of the finite Coxeter group $G_{\overline{p},\overline{q},\overline{r}}$.

This theorem partially answers a question of Serre [S], i.e., he asked for the location of the poles of $p_S(t)$. The proof of this theorem uses a theorem of James Cannon, proven after the above was conjectured. Further information about the poles of $p_S(t)$ is given by the following theorem of Cannon.

<u>Theorem</u> (1.11). If G and S are as in (1.10), then $p_S(t)$ has two positive real poles λ and $1/\lambda$. All other poles of $p_S(t)$ lie on the unit circle.

Cannon's main tool is the theory of reciprocal polynomials. Recall, a polynomial with complex coefficients is called <u>reciprocal</u> if for every root λ, $1/\lambda$ is a root of the same multiplicity. In particular, polynomials f whose coefficients are palindromic ($a_i = a_{n-i}$, $n = \deg f$) or anti-palindromic ($a_i = -a_{n-i}$) are reciprocal.

A <u>Salem number</u> α is an algebraic integer whose monic irreducible polynomial over Q, f_α, is reciprocal with at most one root of modulus > 1. Salem has shown [Sa] that if α is a Salem number which is not a root of unity then f has two reciprocal roots, and all other roots lie on the unit circle. Salem studied these polynomials in connection with some problems in Fourier analysis. Theorem (1.11) implies that λ and $1/\lambda$ are Salem numbers. Cannon has also proven Theorem (1.11) for G = the fundamental group of a compact Riemann surface of genus g and

$$S = \{a_1^{\pm 1}, b_1^{\pm 1}, \ldots, a_g^{\pm 1}, b_g^{\pm 1}\}$$

a set of $4g$ generators for G, then $p_S(t)$ satisfies the conclusion of (1.11).

§2. Growth functions through history

(2.1) The growth function of a group G may be interesting even for a finite group. Suppose W is a finite Coxeter group and let S be the set of all reflections in W. Then

$$p_S(t) = \prod_{i=0}^{\ell} (1 + m_i t)$$

where the m_i are certain well known numbers called the exponents of W. The exponents of a finite Coxeter group can be defined as follows: [B, Ch.V, §6, no.2]. Let g be the Coxeter element of W and h = the order of g. Then there are integers $0 < m_1 < \ldots < m_\ell < h$ so that the eigenvalues of g are $\exp(2\pi i m_j/h)$, $j = 1,\ldots,\ell$. The exponents are also related to the invariant polynomials of W as follows. Every Coxeter group has a canonical real representation. If the graph has ℓ vertices then the group W acts on R^ℓ [B, Ch.V, §4]. This representation induces an action of W on the polynomial ring in ℓ-variables. The ring of invariant polynomials (if W is finite) is isomorphic to a polynomial ring in ℓ-variables. The degrees of the generators of the ring of invariants are $m_1 + 1, m_2 + 1, \ldots, m_\ell + 1$. [Ch], [Co], [St], [B, Ch.V, §5,6].

There is another interesting power series which arises in this context.

<u>Definition</u> (2.2): If X is a topological space and F is a field so that $\dim_F H^i(X,F)$ is finite for all i we can define the Poincare power series of X (relative to F) by

$$p_X(t) = \sum_{i=0}^{\infty} \dim_F H^i(X,F) t^i .$$

If p_X is a polynomial we call it the Poincare polynomial of X.

<u>Example</u> (2.3). Orlik and Solomon [O-S1] have discovered that the growth power series (actually a polynomial) defined above for a finite Coxeter group is actually the Poincare polynomial of $X = C^\ell - \bigcup_{s\in S} H_s$ where the H_s are the reflecting hyperplanes of W, i.e., H_s is the fixed point set of $s \in S$. Note that S is the set of all reflections in W. Orlik and Solomon have generalized their results to

unitary reflection groups. A unitary reflection is a complex $\ell \times \ell$ matrix having 1 as an eigenvalue of multiplicity $\ell - 1$. The corresponding reflecting hyperplane is the eigenspace of the eigenvalue 1. If W is a finite group generated by unitary reflections and $X = C^{\ell} - \bigcup_{s \in S} H_s$ as above then they show that the Poincare power series of X factors

$$p_X(t) = \prod_{i=0}^{\ell} (1 + n_i t) .$$

The n_i above agree with the exponents m_i if G is a real reflection group, but in general are different from the m_i. Terao has shown the above factorization holds for certain arrangements of hyperplanes in C^{ℓ} (called free) which need not arise from groups.

Returning to the finite Coxeter groups, there is another natural choice of generating set S. Namely let $S = \{s_1, \ldots, s_n\}$ where the s_i are the generators corresponding to the vertices of the Coxeter graph Γ. Then Solomon [So] has shown that

$$p_S(t) = \prod_{i=0}^{\ell} (1 + \ldots + t^{m_i})$$

There is a third way of constructing an interesting rational function. Suppose F is a field and R is a graded F-algebra, i.e., R is an F-algebra so that $R = \bigoplus_{i=-\infty}^{\infty} R_i$ where the R_i are F-subspaces of R so that $R_i \cdot R_j \subset R_{i+j}$ for all i,j.

<u>Definition</u> (2.5). Suppose $\dim R_i < \infty$, for all i. Then define the Poincare power series of R to be

$$p_R(t) = \sum_{-\infty}^{\infty} (\dim_F R_i) t^i .$$

The Poincare power series of a topological space is actually a special case of this (obtained by letting $R = H^*(X,F)$ and $R_i = H^i(X,F)$). If R is a finitely generated positively graded (i.e., $R_i = 0$ for $i < 0$), F-algebra then one can show that $P_R(t)$ is a rational function [A-M, Ch.10].

We noted above (2.1), that if W is a finite Coxeter group, there is a natural action of W on the polynomial ring R in ℓ variables and the ring of invariant polynomials $A = R^W$ is isomorphic to a polynomial ring in ℓ variable. Moreover A is generated by polynomials $P_1, \ldots, P_{\ell}$ so that degree $P_i = m_i + 1$. One can easily see that the Poincare power series of the graded ring A is

$$P_A(t) = \prod_{i=1}^{\ell} 1/(1 - t^{m_i+1}) .$$

Bott [Bot] remarks that $R/(P_1,\dots,P_\ell)$ is isomorphic to $H^*(G/B)$ where G is the Lie group associated to W and B is a Borel subgroup. It is interesting to note that the Poincare polynomial of this graded algebra is equal to

$$\prod_{i=0}^{\infty} (1 + \dots + t^{m_i})$$

Example (2.6). Suppose G is the dihedral group of order $2n$. Then G is the Coxeter group corresponding to the Coxeter graph

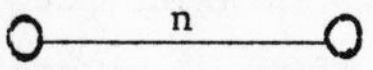

The Coxeter representation of G is as the group generated by the reflections s_1, s_2 in two lines L_1, L_2 so that the angle between L_1 and L_2 is $2\pi/n$. If $S = \{s_1,s_2\}$ one can easily check that

$$p_S(t) = 1 + 2t + 2t^2 + \dots + 2t^{n-1} + t^n$$
$$= (1+t)(1+\dots+t^{n-1}) \qquad \text{if } n \text{ is finite.}$$
$$p_S(t) = (1+t)/(1-t) \qquad \text{if } n = \infty$$

while if S = the set of all reflections and W is finite then

$$p_S(t) = 1 + nt + (n-1)t^2 = (1+t)(1+(n-1)t) .$$

(2.7) The first interesting calculation of a growth series for an infinite group is the generalization of Solomon's result (2.4) to the affine Weyl groups [B, Ch.VI, §4, ex.10], [Bot], [I-M]. If W_a is an affine Weyl group, W is the corresponding finite Weyl group, and $m_1,\dots,m_\ell$ are the exponents of W and S is the set of generators for W corresponding to the vertices of the Coxeter graph then

$$p_S = \prod_{i=1}^{\ell} (1 + \dots + t^{m_i})/(1-t^{m_i}) .$$

Bott's proof is interesting since he relates p_S to the Poincare power series of a topological space, namely the loop space $\Omega(G)$ of the simply connected compact Lie group corresponding to W.

(2.8) Rationality of growth series.

The growth series of a finitely presented group need not be a rational function. Cannon [Cal] has shown that if S is a set of generators for G yielding a finite presentation and the a_i are algorithmically calculable (for example if $p_S(t)$ is a rational function, see 3.8) then the group G has solvable word problem. It is known that there exist groups as above with non-solvable word problem.

Coxeter groups have rational growth series. Bourbaki [B] gives an algorithm for computing p_S for the standard generators. If Γ is a Coxeter graph we let p_Γ denote the growth series of W_Γ relative to S_Γ. Then

$$1/p_\Gamma(t) = -\sum_{\Gamma' \subset \Gamma} \varepsilon(\Gamma')/p_{\Gamma'}(t) \qquad \text{if } W_\Gamma \text{ is infinite.} \qquad (2.8.1)$$

$$\sum_{\Gamma' \subset \Gamma} \varepsilon(\Gamma') p_\Gamma(t)/p_{\Gamma'}(t) = t^m \qquad \text{if } W_\Gamma \text{ is finite.} \qquad (2.8.2)$$

where m = length of the (unique) element of maximal length, $\varepsilon(\Gamma') = (-1)^{\text{card } \Gamma'}$ and Γ' ranges over all subgraphs of Γ. Note that $p_\phi(t) = 1$, where ϕ denotes the empty graph.

These formulas allow one to calculate, in principle, p_Γ by induction. Using these formulas one can see that the zeros of p_Γ are roots of unity, $1/p_\Gamma(\infty)$ is an integer and $1/p_\Gamma(t^{-1}) \in Z[[t]]$.

The formulas above allows one to easily calculate p_Γ for the graph

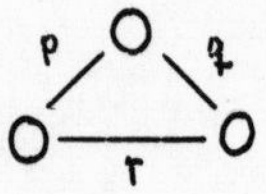

(2.8.3)

to get (1.5). If the expression for p_S in (1.5) is rewritten in a suitable form it is valid even if p, q or $r = \infty$. Namely, let $[n] = 1 + t + \ldots + t^{n-1} = (1-t^n)/(1-t)$ if p is finite and $[\infty] = 1 + t + t^2 + \ldots = 1/(1-t)$. If Γ =

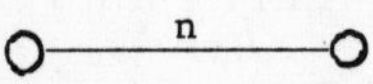

then $p_\Gamma(t) = [2][n]$ (see 2.6). Applying the formula (2.8.1) above we see that if Γ is as in (2.8.1) with $p,q,r \in N \cup \{\infty\}$ and $1/p + 1/q + 1/r \leq 1$ (this is the case when W_Γ is infinite) then

$$p_\Gamma = \frac{[2][p][q][r]}{[2][p][q][r] - t([p-1][q][r] + [p][q-1][r] + [p][q][r-1])}$$

One can verify that the denominator is a reciprocal polynomial if and only if p, q and r are finite.

(2.9) Growth and curvature.

Much of the interest in growth functions was started by Milnor who showed that there is a relation between the curvature of a compact manifold and the growth rate of its fundamental group. In [Mi 1] he proved the following theorems:

Theorem (2.9.1): If M is a complete n-dimensional Riemannian manifold whose mean curvature tensor R_{ij} is everywhere positive semidefinite then the growth function $\{a_i\}$ associated to any finitely generated subgroup of the fundamental group (and any set of generators) must satisfy:

there exists a constant C so that, $a_i < C \cdot i^{n-1}$ for all i.

Definition (2.9.2): If the inequality in (2.9.1) holds for a set of generators of G then it holds for all generating sets [Mi 1] and we say G has polynomial growth.

Theorem (2.9.3): If M is compact Riemannian with all sectional curvatures less than zero, then the growth function of the fundamental group $\pi_1(M)$ is exponential, i.e.,

$$a_i > c^i, \quad \text{for all } i$$

for some constant $c > 1$.

Definition (2.9.4): If a_i satisfies the condition above for one (and hence for all) sets of generators of a group G we say that G has exponential growth.

The study of exponential, non-exponential and polynomial growth was continued by Wolf, Milnor and Bass [Wo], [Mi 1], [Ba]. Milnor showed by a simple direct argument that a finitely generated solvable group not of exponential growth is polycyclic. Wolf showed that if G is polycyclic then it is virtually nilpotent, i.e., G has a nilpotent subgroup of finite index. Bass found an expression for the degree of polynomial growth of a nilpotent group and conjectured that a finitely generated group not of exponential growth is virtually nilpotent. Recently Gromov [G] proved that a finitely generated group of polynomial growth is virtually nilpotent.

§3. Calculation of p_S: some examples.

In this section we will first give some examples for which p_S can be computed directly by counting group elements. The second part of this section will be devoted to showing how geometry can be used to calculate growth series for some Fuchsian groups.

Example (3.1): Let $G = \mathbb{Z}$, $S = \{+1,-1\}$. Then $a_n = 2$ for all $n > 1$ and hence

$$p_S(t) = (1+t)/(1-t)$$

Example (3.2): [Cal] (due to P. Melvin). $G = \mathbb{Z}$, $S = \{\pm 2, \pm 3\}$. Then

$$p_S = 1 + 4t + 8t^2 + \Sigma 6t^n = -5 - 2t + 2t^2 + 6/(1-t) .$$

Melvin has shown that every generating set for Z gives a growth sequence which is eventually constant hence

$$p_S(t) = f(t) + c/(1-t)$$

where f is a polynomial and c is a positive integer.

Proposition (3.3): Suppose $f(t) = \Sigma a_n t^n$, a_n non-negative integers with $a_0 = 1$. If $f(t)$ is a rational function then

(i) a_n has exponential growth if and only if f has a pole at some ξ such that $0 < |\xi| < 1$.

(ii) a_n has polynomial growth if and only if all poles of f are roots of unity.

Proof: [Cal, Theorem 8.5]: By [P-Sz, p.141-144] we can write

$$f(t) = g(t)/h(t)$$

where g and h are relatively prime polynomials with integer coefficients and $h(0) = 1$. It is sufficient to consider the case degree $h = m > 0$. Now $\lim_{n\to\infty} a_n^{1/n}$ exists [Mi 1] and equals $1/R$ where R is the radius of convergence. Thus f has a pole at some ξ inside the unit circle, if and only if a_n has exponential growth. On the other hand, if a_n does not have exponential growth then $R \geq 1$. Now if we let $h(t) = a_m t^m + \ldots + a_1 t + 1$ then the product of the roots of h is $1/a_m$. But all roots have modulus ≥ 1, hence $a_m = 1$ and all roots lie on the unit circle. A polynomial with integer coefficients all of whose roots have modulus 1 is a product of cyclotomic polynomials (by Kronecker's theorem). Thus a_n has polynomial growth implies a_n does not have exponential growth which implies all poles of f are roots of unity. Conversely one can easily show that if the poles of f are roots of unity then a_n has polynomial growth.

<u>Proposition</u> (3.4): $G = H \times K$ is a semi direct product, S_H and S_K are generating sets for H and K respectively, $kS_Hk^{-1} = S_H$ for all $k \in K$ and

$$S = S_H \cup S_K$$

then

$$p_S(t) = p_{S_H}(t) \cdot p_{S_K}(t) .$$

<u>Corollary</u>: If G is the free abelian group on generators $e_1,\ldots,e_n$ and $S = \{\pm e_1,\ldots,\pm e_n\}$ then

$$p_S(t) = ((1+t)/(1-t))^n .$$

<u>Example</u> (3.5): If G is the free non-abelian group on n-generators $e_1,\ldots,e_n$ and $S = \{e^{\pm 1}_1,\ldots,e^{\pm 1}_n\}$ then

$$p_S(t) = 1 + \sum_{i=1}^{\infty} (2n)(2n-1)^{i-1}t^i = (1+t)/(1-(2n-1)t)$$

Proof: We prove that $a_i = (2n)(2n-1)^{i-1}$ by induction on i. Clearly, $a_1 = 2n$. Assume the assertion true for $i-1$. Every element of length i has a unique representation in the form $g = \varepsilon_1 \ldots \varepsilon_i$ where $\varepsilon_j \in S$ for all j and $\varepsilon_j \neq \varepsilon^{-1}_{j+1}$ for all $j = 1,\ldots,i-1$. Thus g can be written uniquely in the form $g's$ where

g' has length $i-1$ and s is not the inverse of the last element in g'. So for each g' there are $2n-1$ possible choices for $s \in S$. Thus

$$a_i = (2n-1)a_{i-1} = (2n-1)(2n)(2n-1)^{i-2}$$

by the inductive hypothesis. This is the desired result.

(3.6) Change of generating set.

If S and T are two generating sets for G, it seems difficult in general to see the relation between p_S and p_T. There is a rather weak statement that one can make when S is a subset of T. We shall also give a special case where one can give precise information.

The statement of the result is more natural when given in terms of the "cumulative growth series of G". To be precise let

$$q_S(t) = \Sigma\, b_n t^n$$

where $b_n = \{g \in G \mid \ell_S(g) \leq n\}$. Thus

$$b_n = \sum_{i=0}^{n} a_i$$

and hence

$$q_S(t) = p_S(t)/(1-t) .$$

Proposition: If $S \subset T$ then $q_T(t) - q_S(t)$ has non-negative coefficients.

Proof: If $\ell_S(g) \leq n$ then clearly $\ell_T(g) \leq n$.

(3.7): If we take T to be the set of all elements of G which are words of length $\leq k$ for some k, then we can calculate p_T from p_S.

Proposition: Suppose S is a generating set for G, $k > 0$ and

$$T = \{g \in G \mid \ell_S(g) \leq k\} .$$

Let $c_i = \#\{g \in G \mid \ell_T(g) \leq i\}$ and $b_i = \{g \in G \mid \ell_S(g) \leq i\}$. Then

$$c_i = b_{ki}, \quad \text{for } i > 0 .$$

Proof: One can easily verify that $\ell_S(g) \leq ki$ if and only if $\ell_T(g) \leq i$.

Corollary: $k \cdot p_T(t^k) = p_S(t) + p_S(\omega t) + \ldots + p_S(\omega^{k-1} t)$ where ω is a primitive k^{th} root of 1.

(3.8). Calculations for Fuchsian Groups.

Cannon's method for computing p_S involves looking at the Cayley graph associated to a presentation of a group. We show that there is an alternative method of calculating p_S when the group is a discrete transformation group and the generating set has geometric significance. This method involves looking at the tesselation defined by the translates of a fundamental region for the action.

In order to motivate what follows we first discuss the significance of the rationality of p_S. Let $p(t) = \Sigma a_n t^n$ denote any power series with integer coefficients. It is shown in [P-Sz, p.141-142] that if p is a rational function then we can write

$$p(t) = f(t)/g(t)$$

where g and f are polynomials with integer coefficients. Moreover, if $a_0 = 1$ we can assume the constant terms of g and f are 1. Suppose that $g(t) = \sum_{i=0}^{m} b_i t^i$ and $f(t) = \sum_{i=0}^{n} c_i t^i$. Then we have a linear recursion relation for the a_i, $i > n$

$$a_i = -(b_1 a_{i-1} + \ldots + b_m a_{i-m}) \qquad (3.8.1)$$

For $i \leq n$ we have

$$a_i = -(b_1 a_{i-1} + \ldots + b_m a_{i-m}) + c_i \qquad (3.8.2)$$

(define $a_i = 0$ for $i < 0$).

These equations come from comparing the coefficients of t^i in the identity $p(t)g(t) = f(t)$. Conversely, if we are given a sequence of a_i, $i \geq 0$, which are defined by a linear recurrence relation as in (3.8.1), (3.8.2) then we can easily read off f and g. For example if $\{a_i\}$ is the Fibonacci sequence

$$a_0 = a$$

$$a_1 = b$$

$$a_i = a_{i-1} + a_{i-2} \qquad i \geq 2$$

then

$$p(t) = (a + (b-a)t)/(1 - t - t^2)$$

In summary, to calculate p_S it is sufficient to find a linear recurrence relation for $\{a_i\}$.

(3.9). Tesselations.

Now we specialize to the case where G is a discrete group of isometries of the upper half-plane H_+ or Euclidean plane $\mathbb{R}^2$. We assume that G has a fundamental region F_0 such that

(i) area F_0 is finite.

(ii) F_0 has a finite number of sides each of which is a hyperbolic line segment.

The translates of F_0 under G cover H_+ and $g(\dot{F}_0) \cap h(\dot{F}_0) \neq \phi$ implies $g = h$

(where $\dot{F}$ denotes interior). Thus the translates of F_0 give a tesselation of the plane.

Let $\theta=\{g(\overline{F}_0) \mid g \in G\}$ (where $\overline{F}_0$ denotes the closure of F_0). An element $F \in \theta$ will be called a tile of the tesselation. We can define the length of a tile inductively, as follows:

(1) $\ell(F_0) = 0$.

(2) Suppose the tiles of length $\leq n-1$ have been defined. Then F is a tile of length n if its length has not yet been defined and F is adjacent to a tile of length $n-1$.

Clearly, every tile of the tesselation θ has a well defined length.

(3.10). It is not hard to see that this notion of length of a tesselation is related to a length function on G for a suitable generating set S_{F_0}, which is defined as follows:

If F_0 is a fundamental region as above then for each side Σ of F_0 there is an element $g_\Sigma \in F_0$ such that either

(1) g_Σ identifies Σ with another side of F_0 or

(2) g_Σ is the reflection in Σ.

Let $S_{F_0} = \{g_\Sigma \mid \Sigma$ is a side of $F_0\}$. Then S_F is a generating set for G and

$$\ell_S(g) = \ell(g(\overline{F}_0)) \quad \text{for all } g \in G.$$

Definition (3.11): (The growth function of a tesselation). If we have a tesselation as above let

$$a_n = \#\{F \in \theta \mid \ell(F) = n\}$$

the number of tiles of weight n. Then we can define the growth power series of θ to be

$$p_\theta(t) = \Sigma\, a_n t^n$$

By the remark above if $S = S_{F_0}$ then

$$p_S(t) = p_\theta(t) .$$

Note: This definition can be made for any tesselation of a topological space. It does depend on the choice of a base tile F_0. Terao and the author have studied some examples of finite tesselations associated to free arrangements of hyperplanes (in the sense of Terao [T]). The associated growth power series (polynomials in this case) appear to have nice factorization properties (cf. 2.4) provided the 'correct' base tile is chosen.

We will be looking for recurrence relations defining the sequences $\{a_n\}$. To do this it appears to be necessary to introduce some auxiliary notions, the weight of a vertex and "overlapping tiles".

Definition (3.12): A tile of weight n is said to be overlapping if it is adjacent to at least 2 tiles of weight $n-1$. We let b_n be the number of overlapping tiles of weight n.

Definition (3.13): The weight of a vertex v is the smallest n so that v is a vertex of a tile F of weight n. Let c_n be the number of vertices of weight n. Similarly one can define the weight of an edge Σ as the smallest n so that Σ is an edge of a tile of weight n.

Example (3.14): If we let θ be the tesselation of the Euclidean plane by equilateral triangles we get

n	a_n	b_n	c_n
0	1	0	3
1	3	0	3
2	6	0	6
3	8	2	6

Our first calculation of p_S will be for the covering group of a compact Riemann Surface. This was first done by Cannon [Cal] using a different method.

Proposition (3.15): Suppose G is the covering group of a compact Riemann surface of genus $g \geq 2$ and F_0 is a $4g$ sided fundamental region for G, then

$$p_{F_0}(t) = \frac{(1+t)(1-t^{2g})}{1-(4g-1)t+(4g-1)t^{2g}-t^{2g+1}}$$

Thus if S is the corresponding set of $4g$ generators for G we see that $p_S(t)$ is as above.

Proof: The proposition depends on the following lemma.

Lemma (3.15.1):

$$a_n = (4g-1)a_{n-1} - b_{n-1} - b_n \quad , \text{ for } n \geq 2 \qquad (1)$$

$$b_n = c_{n-2g} \quad , \text{ for } n \geq 0 \qquad (2)$$

$$c_n = (4g-2)a_n - b_n \quad , \text{ for } n \geq 1 \qquad (3)$$

Equation (2) makes sense for all $n < 0$ if we define $c_i = 0$ for $i < 0$. If we let $b(t) = \Sigma\, b_n t^n$ and $c(t) = \Sigma\, c_n t^n$ then

$$p(t) = (4g-1)t\, p(t) - t\, b(t) - b(t) + 1 + t \qquad (1^*)$$

$$b(t) = t^{2g} c(t) \qquad (2^*)$$

$$c(t) = (4g-2)p(t) - b(t) + 2 \qquad (3^*)$$

Proof of the proposition: (1*), (2*), and (3*) are 3 linear equations in 3 unknowns, p(t), b(t), c(t). Solve for p(t) by substituting for b(t) using (2*). We get

$$(1-(4g-1)t)p(t) + (1+t)t^{2g}c(t) = 1 + t$$

$$(2-4g)p(t) + (1+t^{2g})c(t) = 2$$

and solving for p gives the desired result.

Proof of the lemma: The equations (1*), (2*), and (3*) follow immediately from (1), (2) and (3), respectively and a direct check of the coefficients of low degree. To verify (1), (2), and (3) we first note that each vertex of the tesselation lies on 4g tiles. It is easy to see that $a_0 = 1$, $b_0 = 1$, $c_0 = 4g$, $a_1 = 4g$, $b_1 = 0$, $c_1 = (4g-2)a_1$. If v is a vertex of weight i then v lies on 2 tiles of weight i+1, 2 tiles of weight i+2,...,2 tiles of weight i+2g-1 and one tile of weight i+2g.

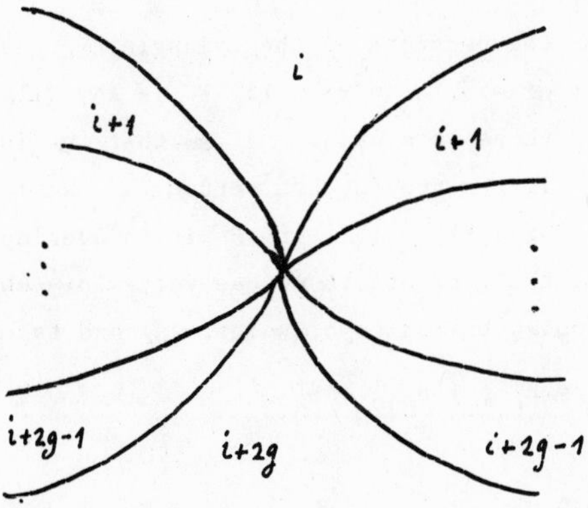

For each vertex of weight i there is a unique overlapping face of weight i+2g. This proves (2). To prove (1) we note that every tile of weight n is adjacent to at least one and at most two tiles of weight n-1. A tile of weight n-1 will have (4g-1) edges of weight n-1 if it is non-overlapping and (4g-2) edges of weight n-1 if it is overlapping. Now # of tiles of weight n = # edges of weight n-1 - # overlapping tiles of weight n, hence

$$a_n = ((4g-1)a_{n-1} - b_{n-1}) - b_n$$

which gives us (1). Finally, to compute c_n we note that a tile of weight n has 4g-2 vertices of weight n if it is not overlapping and 4g-3 vertices of weight n if it is overlapping. Every vertex of weight n lies on a unique tile of weight n, hence (3) follows.

(3.16). We shall now apply the above methods to calculate the growth series of a triangle group. Of course, this can be done using the method of [B] since a triangle group is also a Coxeter group. However, our method is applicable in the non-Coxeter case. Suppose F_0 is a triangle with angles π/p, π/q and π/r and assume $1/p + 1/q + 1/r \leq 1$. If $1/p + 1/q + 1/r = 1$ then our triangle is Euclidean, otherwise it is hyperbolic. Recall that $G = \Delta_{p,q,r}$ is the group generated by the reflections in the sides of F_0.

Proposition (3.16.1): If G is as above and $S = \{s_1,s_2,s_3\}$ are the three reflections in the sides of F_0, then S generates G and

$$p_S(t) = \frac{(1+t)(1-t^p)(1-t^q)(1-t^r)}{1 - 2t + t^{p+1} + t^{q+1} + t^{r+1} - t^{p+q} - t^{p+r} - t^{q+r} + 2t^{p+q+r} - t^{p+q+r+1}}$$

In order to prove the proposition we must make some definitions and prove a lemma.

Definition (3.16.2): Number the vertices of the triangle F_0 so that the angle at v_i is π/e_i where $e_1 = p$, $e_2 = q$, $e_3 = r$. If F is any tile of the tesselation and $v \in F$ is a vertex, there is a unique i so that v is the image of v_i for some $g \in G$. We say i is the type of the vertex v. We let $c_{n,i}$ = the number of vertices of weight n and type i. If F is an overlapping triangle we define the type of F to be the type of its unique vertex of lowest weight. Let $b_{n,i}$ = the number of overlapping triangles of weight n and type i. Clearly

$$b_n = b_{n,1} + b_{n,2} + b_{n,3}$$

and

$$c_n = c_{n,1} + c_{n,2} + c_{n,3}$$

Let $\beta_i(t) = \Sigma\, b_{n,i} t^n$,

$\beta(t) = \beta_1(t) + \beta_2(t) + \beta_3(t)$,

$\gamma_i(t) = \Sigma\, c_{n,i} t^n$.

Lemma (3.16.3): If G and S are as above then

$$a_n = 2a_{n-1} - b_n - b_{n-1}, \qquad \text{for } n \geq 2 \qquad (1)$$

$$b_{n,i} = c_{n-e_i,i}, \qquad \text{for } n \geq 0 \qquad (2)$$

$$c_{n,1} = c_{n-1,2} + c_{n-1,3} + b_{n-1,1} - b_{n,2} - b_{n,3}, \qquad \text{for } n \geq 2$$

$$c_{n,2} = c_{n-1,1} + c_{n-1,3} + b_{n-1,2} - b_{n,1} - b_{n,3}, \qquad \text{for } n \geq 2 \qquad (3)$$

$$c_{n,3} = c_{n-1,1} + c_{n-1,2} + b_{n-1,3} - b_{n,1} - b_{n,2}, \qquad \text{for } n \geq 2$$

and

$$p_S(t) = 2tp_S(t) - \beta(t) - t\beta(t) + 1 + t \tag{1*}$$

$$\beta_i(t) = t^{e_i}\gamma_i(t) \quad \text{for } i = 1,2,3 \tag{2*}$$

$$\gamma_1(t) = t\gamma_2(t) + t\gamma_3(t) + t\beta_1(t) - \beta_2(t) - \beta_3(t) + 1 - t$$

$$\gamma_2(t) = t\gamma_1(t) + t\gamma_3(t) + t\beta_2(t) - \beta_1(t) - \beta_3(t) + 1 - t \tag{3*}$$

$$\gamma_3(t) = t\gamma_1(t) + t\gamma_2(t) - t\beta_3(t) - \beta_1(t) - \beta_2(t) + 1 - t$$

<u>Proof of proposition:</u> Substitute for the β_i in (1*) and (3*) using (2*). Then we get 4 linear equations in 4 unknowns $p, \gamma_1, \gamma_2, \gamma_3$. The matrix of the equation is

$$\begin{matrix} 1-2t & (1+t)t^{e_1} & (1+t)t^{e_2} & (1+t)t^{e_3} & 1+t \\ 0 & 1-t^{e_1+1} & -t+t^{e_2} & -t+t^{e_3} & 1-t \\ 0 & -t+t^{e_1} & 1-t^{e_2+1} & -t+t^{e_3} & 1-t \\ 0 & -t+t^{e_1} & -t+t^{e_2} & 1-t^{e_3+1} & 1-t \end{matrix}$$

Solve for p using Cramer's rule. For the numerator we get

$$(1+t)^3(1-2t)(1-t^{e_1})(1-t^{e_2})(1-t^{e_3}) .$$

Calculate the denominator and cancel $(1+t)^2(1-2t)$ and we get the desired result.

<u>Proof of Lemma</u> (3.16.3); (1*), (2*) and (3*) follow from (1), (2) and (3) and direct calculation of the coefficients for $n = 0$ and 1. Equations (1) and (2) are proven the same way as in Lemma (3.15.1). Finally, we prove the first equation of (3) (the others follow by symmetry). Suppose $n \geq 2$. Let A = set of all vertices of type 1 and weight n. At each vertex v of weight i in the tesselation we have the weights of the tiles meeting v are as follows:

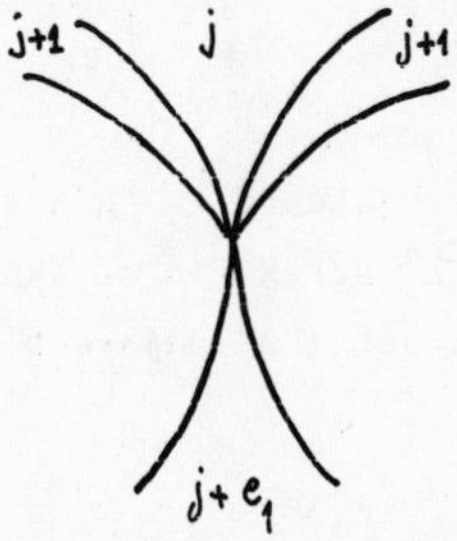

since there is an even number of angles at v. If v is a vertex let F_v be the unique triangle so that $v \in F_v$ and weight v = weight F_v. Note that F_v cannot be an overlapping tile (since in that case weight $v <$ weight F for all vertices of F). Thus F_v is adjacent to a unique tile of weight $n-1$. Let

$$B = \{v \in A \mid F_v \text{ is adjacent to an overlapping triangle}\} .$$

If $v \in B$ we have the following diagram

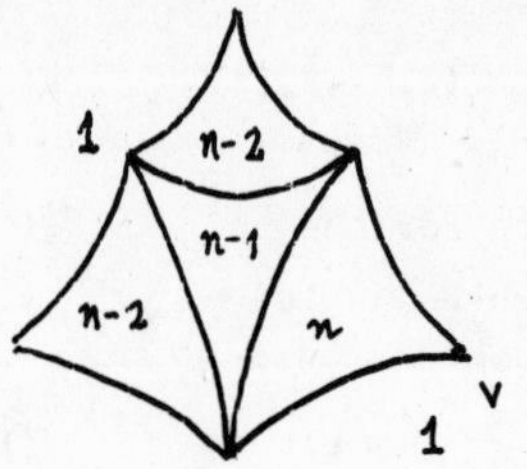

The overlapping triangle must be type 1, hence $\#B = b_{n-1,1}$. On the other hand, if $v \in A - B$ then we have

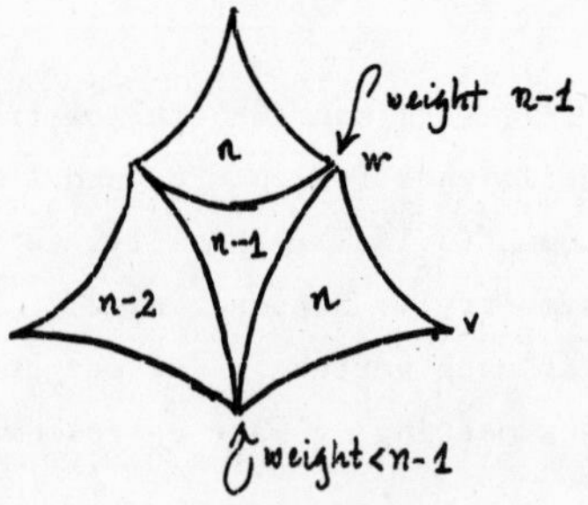

Thus for each $v \in A - B$ we can associate to v a unique vertex w of weight $n-1$ and type $\neq 1$. Conversely, given a vertex w of type $\neq 1$ and weight $n-1$ there is a unique vertex of type 1 and weight n adjacent to w unless the tile F is overlapping. Thus

$$\#(A-B) = c_{n,2} + c_{n,3} - b_{n,2} - b_{n,3}$$

$$\#B = b_{n-1,1} .$$

This gives the desired result.

BIBLIOGRAPHY

[A-M] M. Atiyah and I. MacDonald, Commutative Algebra, Addison-Wesley, Reading Massachusetts.

[Ba] H. Bass, The degree of polynomial growth of finitely generated groups, Proc. London Math. Soc. (3) 25 (1972), 603-614.

[B] N. Bourbaki, Groupes et algebres de Lie, Chapters 4, 5 et 6. Paris, Hermann, 1968.

[Bor] A. Borel, Sur la cohomologie des espaces fibres princepaux et des espaces homogenes de groupes de Lie compacts, Annals of Math., 57 (1956), 115-207.

[Bot] M.R. Bott, An application of the Morse theory to the topology of Lie groups, Bull. Soc. Math. France, 84 (1956), 251-281.

[Ca1] J. Cannon, The growth of the closed surface groups and compact hyperbolic Coxeter groups (preprint).

[Ca2] J. Cannon, The combinatorial structure of co-compact discrete hyperbolic groups, (preprint).

[Car] P. Cartier, Les arrangements d'hyperplans: un chapitre de geometrie combinatoire. Sem. Bourbaki 1980/81, no. 561.

[Ch] C. Chevalley, Invariants of finite groups generated by reflections, Amer. J. Math., 77 (1955), 778-782.

[Co] A.J. Coleman, The Betti numbers of the simple groups, Can. J. Math. 10 (1958), 349-356.

[Cox] H.S.M. Coxeter, The product of generators of a finite group generated by reflections, Duke Math. Jour., 43 (1951), 765-782.

[C-M] H.S.M. Coxeter and W.O. Moser, Generators and relators for discrete groups, New York, Springer.

[D] A. Durfee, Fourteen characterizations of rational double points and simple critical points.

[G] D. Gromov, Groups of polynomial growth and expanding maps, I.H.E.S., 1980 (preprint).

[I-M] M. Iwahori and H. Matsumoto, On some Bruhat decomposition and the structure of Hecke rings of p-adic Chevalley groups, Publ. Math. Inst. Hautes Et. Sci., 25 (1965), 5-48.

[Mi 1] J. Milnor, A note on curvature and the fundamental group, J. Diff. Geometry, 2 (1968), 1-7.

[Mi 2] J. Milnor, Growth of finitely generated solvable groups, J. Diff. Geometry, 2 (1968), 447-449.

[O-S1] P. Orlik and L. Solomon, Combinatorics and topology of complements of hyperplanes, Invent. Math. 56 (1980), 167-189.

[O-S2] P. Orlik and L. Solomon, Unitary reflection groups and cohomology, Inv. Math., 59 (1980), 77-94.

[P-S] G. Polya and Szego, Aufgaben und lehrsatze aus der analysis II, Berlin, Springer, 1925.

[Sa] R. Salem, Power series with integer coefficients, Duke Math. J., 12 (1945), 153-172.

[Se] J.P. Serre, Cohomologie des groupes discrets, in Prospects in Mathematics, Princeton, Princeton University Press, (1971), 77-169 (esp. p. 112 remark and p. 145 prop/ 26(d)).

[So] L. Solomon, The orders of the finite Chevalley groups, J. of Algebra, 3 (1966), 376-393.

[St] R. Steinberg, Finite reflection groups, Trans. Amer. Math. Soc., 91 (1959), 493-504.

[Sv] A.S. Svarc, A volume invariant of coverings, Dokl. Akad. Nauk SSSR, 105 (1953), 32-34.

[T] H. Terao, Generalized exponents of a free arrangement of hyperplanes and Shepherd-Todd-Brieskorn formula, Invent. Math.

[Ti] J. Tits, Groupes a croissance polynomiale, Sem. Bourbaki, 1980/81, no. 572.

[W] P. Wagreich, Growth functions of discrete groups in dimension 2, to appear.

[Wo] J.A. Wolf, Growth of finitely generated solvable groups and curvature of Riemannian manifolds, J. Diff. Geometry, 2 (1968), 421-446.

Department of Mathematics
University of Illinois at Chicago Circle
Box 4348
Chicago, Illinois 60680

Vol. 817: L. Gerritzen, M. van der Put, Schottky Groups and Mumford Curves. VIII, 317 pages. 1980.

Vol. 818: S. Montgomery, Fixed Rings of Finite Automorphism Groups of Associative Rings. VII, 126 pages. 1980.

Vol. 819: Global Theory of Dynamical Systems. Proceedings, 1979. Edited by Z. Nitecki and C. Robinson. IX, 499 pages. 1980.

Vol. 820: W. Abikoff, The Real Analytic Theory of Teichmüller Space. VII, 144 pages. 1980.

Vol. 821: Statistique non Paramétrique Asymptotique. Proceedings, 1979. Edited by J.-P. Raoult. VII, 175 pages. 1980.

Vol. 822: Séminaire Pierre Lelong–Henri Skoda, (Analyse) Années 1978/79. Proceedings. Edited by P. Lelong et H. Skoda. VIII, 356 pages, 1980.

Vol. 823: J. Král, Integral Operators in Potential Theory. III, 171 pages. 1980.

Vol. 824: D. Frank Hsu, Cyclic Neofields and Combinatorial Designs. VI, 230 pages. 1980.

Vol. 825: Ring Theory, Antwerp 1980. Proceedings. Edited by F. van Oystaeyen. VII, 209 pages. 1980.

Vol. 826: Ph. G. Ciarlet et P. Rabier, Les Equations de von Kármán. VI, 181 pages. 1980.

Vol. 827: Ordinary and Partial Differential Equations. Proceedings, 1978. Edited by W. N. Everitt. XVI, 271 pages. 1980.

Vol. 828: Probability Theory on Vector Spaces II. Proceedings, 1979. Edited by A. Weron. XIII, 324 pages. 1980.

Vol. 829: Combinatorial Mathematics VII. Proceedings, 1979. Edited by R. W. Robinson et al.. X, 256 pages. 1980.

Vol. 830: J. A. Green, Polynomial Representations of GL_n. VI, 118 pages. 1980.

Vol. 831: Representation Theory I. Proceedings, 1979. Edited by V. Dlab and P. Gabriel. XIV, 373 pages. 1980.

Vol. 832: Representation Theory II. Proceedings, 1979. Edited by V. Dlab and P. Gabriel. XIV, 673 pages. 1980.

Vol. 833: Th. Jeulin, Semi-Martingales et Grossissement d'une Filtration. IX, 142 Seiten. 1980.

Vol. 834: Model Theory of Algebra and Arithmetic. Proceedings, 1979. Edited by L. Pacholski, J. Wierzejewski, and A. J. Wilkie. VI, 410 pages. 1980.

Vol. 835: H. Zieschang, E. Vogt and H.-D. Coldewey, Surfaces and Planar Discontinuous Groups. X, 334 pages. 1980.

Vol. 836: Differential Geometrical Methods in Mathematical Physics. Proceedings, 1979. Edited by P. L. García, A. Pérez-Rendón, and J. M. Souriau. XII, 538 pages. 1980.

Vol. 837: J. Meixner, F. W. Schäfke and G. Wolf, Mathieu Functions and Spheroidal Functions and their Mathematical Foundations Further Studies. VII, 126 pages. 1980.

Vol. 838: Global Differential Geometry and Global Analysis. Proceedings 1979. Edited by D. Ferus et al. XI, 299 pages. 1981.

Vol. 839: Cabal Seminar 77 – 79. Proceedings. Edited by A. S. Kechris, D. A. Martin and Y. N. Moschovakis. V, 274 pages. 1981.

Vol. 840: D. Henry, Geometric Theory of Semilinear Parabolic Equations. IV, 348 pages. 1981.

Vol. 841: A. Haraux, Nonlinear Evolution Equations- Global Behaviour of Solutions. XII, 313 pages. 1981.

Vol. 842: Séminaire Bourbaki vol. 1979/80. Exposés 543–560. IV, 317 pages. 1981.

Vol. 843: Functional Analysis, Holomorphy, and Approximation Theory. Proceedings. Edited by S. Machado. VI, 636 pages. 1981.

Vol. 844: Groupe de Brauer. Proceedings. Edited by M. Kervaire and M. Ojanguren. VII, 274 pages. 1981.

Vol. 845: A. Tannenbaum, Invariance and System Theory: Algebraic and Geometric Aspects. X, 161 pages. 1981.

Vol. 846: Ordinary and Partial Differential Equations, Proceedings. Edited by W. N. Everitt and B. D. Sleeman. XIV, 384 pages. 1981.

Vol. 847: U. Koschorke, Vector Fields and Other Vector Bundle Morphisms – A Singularity Approach. IV, 304 pages. 1981.

Vol. 848: Algebra, Carbondale 1980. Proceedings. Ed. by R. K. Amayo. VI, 298 pages. 1981.

Vol. 849: P. Major, Multiple Wiener-Itô Integrals. VII, 127 pages. 1981.

Vol. 850: Séminaire de Probabilités XV. 1979/80. Avec table générale des exposés de 1966/67 à 1978/79. Edited by J. Azéma and M. Yor. IV, 704 pages. 1981.

Vol. 851: Stochastic Integrals. Proceedings, 1980. Edited by D. Williams. IX, 540 pages. 1981.

Vol. 852: L. Schwartz, Geometry and Probability in Banach Spaces. X, 101 pages. 1981.

Vol. 853: N. Boboc, G. Bucur, A. Cornea, Order and Convexity in Potential Theory: H-Cones. IV, 286 pages. 1981.

Vol. 854: Algebraic K-Theory. Evanston 1980. Proceedings. Edited by E. M. Friedlander and M. R. Stein. V, 517 pages. 1981.

Vol. 855: Semigroups. Proceedings 1978. Edited by H. Jürgensen, M. Petrich and H. J. Weinert. V, 221 pages. 1981.

Vol. 856: R. Lascar, Propagation des Singularités des Solutions d'Equations Pseudo-Différentielles à Caractéristiques de Multiplicités Variables. VIII, 237 pages. 1981.

Vol. 857: M. Miyanishi. Non-complete Algebraic Surfaces. XVIII, 244 pages. 1981.

Vol. 858: E. A. Coddington, H. S. V. de Snoo: Regular Boundary Value Problems Associated with Pairs of Ordinary Differential Expressions. V, 225 pages. 1981.

Vol. 859: Logic Year 1979–80. Proceedings. Edited by M. Lerman, J. Schmerl and R. Soare. VIII, 326 pages. 1981.

Vol. 860: Probability in Banach Spaces III. Proceedings, 1980. Edited by A. Beck. VI, 329 pages. 1981.

Vol. 861: Analytical Methods in Probability Theory. Proceedings 1980. Edited by D. Dugué, E. Lukacs, V. K. Rohatgi. X, 183 pages. 1981.

Vol. 862: Algebraic Geometry. Proceedings 1980. Edited by A. Libgober and P. Wagreich. V, 281 pages. 1981.

Vol. 863: Processus Aléatoires à Deux Indices. Proceedings, 1980. Edited by H. Korezlioglu, G. Mazziotto and J. Szpirglas. V, 274 pages. 1981.

Vol. 864: Complex Analysis and Spectral Theory. Proceedings, 1979/80. Edited by V. P. Havin and N. K. Nikol'skii, VI, 480 pages. 1981.

Vol. 865: R. W. Bruggeman, Fourier Coefficients of Automorphic Forms. III, 201 pages. 1981.

Vol. 866: J.-M. Bismut, Mécanique Aléatoire. XVI, 563 pages. 1981.

Vol. 867: Séminaire d'Algèbre Paul Dubreil et Marie-Paule Malliavin. Proceedings, 1980. Edited by M.-P. Malliavin. V, 476 pages. 1981.

Vol. 868: Surfaces Algébriques. Proceedings 1976–78. Edited by J. Giraud, L. Illusie et M. Raynaud. V, 314 pages. 1981.

Vol. 869: A. V. Zelevinsky, Representations of Finite Classical Groups. IV, 184 pages. 1981.

Vol. 870: Shape Theory and Geometric Topology. Proceedings, 1981. Edited by S. Mardešić and J. Segal. V, 265 pages. 1981.

Vol. 871: Continuous Lattices. Proceedings, 1979. Edited by B. Banaschewski and R.-E. Hoffmann. X, 413 pages. 1981.

Vol. 872: Set Theory and Model Theory. Proceedings, 1979. Edited by R. B. Jensen and A. Prestel. V, 174 pages. 1981.

Vol. 873: Constructive Mathematics, Proceedings, 1980. Edited by F. Richman. VII, 347 pages. 1981.

Vol. 874: Abelian Group Theory. Proceedings, 1981. Edited by R. Göbel and E. Walker. XXI, 447 pages. 1981.

Vol. 875: H. Zieschang, Finite Groups of Mapping Classes of Surfaces. VIII, 340 pages. 1981.

Vol. 876: J. P. Bickel, N. El Karoui and M. Yor. Ecole d'Eté de Probabilités de Saint-Flour IX – 1979. Edited by P. L. Hennequin. XI, 280 pages. 1981.

Vol. 877: J. Erven, B.-J. Falkowski, Low Order Cohomology and Applications. VI, 126 pages. 1981.

Vol. 878: Numerical Solution of Nonlinear Equations. Proceedings, 1980. Edited by E. L. Allgower, K. Glashoff, and H.-O. Peitgen. XIV, 440 pages. 1981.

Vol. 879: V. V. Sazonov, Normal Approximation – Some Recent Advances. VII, 105 pages. 1981.

Vol. 880: Non Commutative Harmonic Analysis and Lie Groups. Proceedings, 1980. Edited by J. Carmona and M. Vergne. IV, 553 pages. 1981.

Vol. 881: R. Lutz, M. Goze, Nonstandard Analysis. XIV, 261 pages. 1981.

Vol. 882: Integral Representations and Applications. Proceedings, 1980. Edited by K. Roggenkamp. XII, 479 pages. 1981.

Vol. 883: Cylindric Set Algebras. By L. Henkin, J. D. Monk, A. Tarski, H. Andréka, and I. Németi. VII, 323 pages. 1981.

Vol. 884: Combinatorial Mathematics VIII. Proceedings, 1980. Edited by K. L. McAvaney. XIII, 359 pages. 1981.

Vol. 885: Combinatorics and Graph Theory. Edited by S. B. Rao. Proceedings, 1980. VII, 500 pages. 1981.

Vol. 886: Fixed Point Theory. Proceedings, 1980. Edited by E. Fadell and G. Fournier. XII, 511 pages. 1981.

Vol. 887: F. van Oystaeyen, A. Verschoren, Non-commutative Algebraic Geometry, VI, 404 pages. 1981.

Vol. 888: Padé Approximation and its Applications. Proceedings, 1980. Edited by M. G. de Bruin and H. van Rossum. VI, 383 pages. 1981.

Vol. 889: J. Bourgain, New Classes of $\mathcal{L}^p$-Spaces. V, 143 pages. 1981.

Vol. 890: Model Theory and Arithmetic. Proceedings, 1979/80. Edited by C. Berline, K. McAloon, and J.-P. Ressayre. VI, 306 pages. 1981.

Vol. 891: Logic Symposia, Hakone, 1979, 1980. Proceedings, 1979, 1980. Edited by G. H. Müller, G. Takeuti, and T. Tugué. XI, 394 pages. 1981.

Vol. 892: H. Cajar, Billingsley Dimension in Probability Spaces. III, 106 pages. 1981.

Vol. 893: Geometries and Groups. Proceedings. Edited by M. Aigner and D. Jungnickel. X, 250 pages. 1981.

Vol. 894: Geometry Symposium. Utrecht 1980, Proceedings. Edited by E. Looijenga, D. Siersma, and F. Takens. V, 153 pages. 1981.

Vol. 895: J.A. Hillman, Alexander Ideals of Links. V, 178 pages. 1981.

Vol. 896: B. Angéniol, Familles de Cycles Algébriques – Schéma de Chow. VI, 140 pages. 1981.

Vol. 897: W. Buchholz, S. Feferman, W. Pohlers, W. Sieg, Iterated Inductive Definitions and Subsystems of Analysis: Recent Proof-Theoretical Studies. V, 383 pages. 1981.

Vol. 898: Dynamical Systems and Turbulence, Warwick, 1980. Proceedings. Edited by D. Rand and L.-S. Young. VI, 390 pages. 1981.

Vol. 899: Analytic Number Theory. Proceedings, 1980. Edited by M.I. Knopp. X, 478 pages. 1981.

Vol. 900: P. Deligne, J. S. Milne, A. Ogus, and K.-Y. Shih, Hodge Cycles, Motives, and Shimura Varieties. V, 414 pages. 1982.

Vol. 901: Séminaire Bourbaki vol. 1980/81 Exposés 561–578. III, 299 pages. 1981.

Vol. 902: F. Dumortier, P.R. Rodrigues, and R. Roussarie, Germs of Diffeomorphisms in the Plane. IV, 197 pages. 1981.

Vol. 903: Representations of Algebras. Proceedings, 1980. Editec by M. Auslander and E. Lluis. XV, 371 pages. 1981.

Vol. 904: K. Donner, Extension of Positive Operators and Korovkir Theorems. XII, 182 pages. 1982.

Vol. 905: Differential Geometric Methods in Mathematical Physics Proceedings, 1980. Edited by H.-D. Doebner, S.J. Andersson, an H.R. Petry. VI, 309 pages. 1982.

Vol. 906: Séminaire de Théorie du Potentiel, Paris, No. 6. Proceed ings. Edité par F. Hirsch et G. Mokobodzki. IV, 328 pages. 1982.

Vol. 907: P. Schenzel, Dualisierende Komplexe in der lokale Algebra und Buchsbaum-Ringe. VII, 161 Seiten. 1982.

Vol. 908: Harmonic Analysis. Proceedings, 1981. Edited by F. Ric and G. Weiss. V, 325 pages. 1982.

Vol. 909: Numerical Analysis. Proceedings, 1981. Edited by J.F Hennart. VII, 247 pages. 1982.

Vol. 910: S.S. Abhyankar, Weighted Expansions for Canonical D singularization. VII, 236 pages. 1982.

Vol. 911: O.G. Jørsboe, L. Mejlbro, The Carleson-Hunt Theorem c Fourier Series. IV, 123 pages. 1982.

Vol. 912: Numerical Analysis. Proceedings, 1981. Edited by G. Watson. XIII, 245 pages. 1982.

Vol. 913: O. Tammi, Extremum Problems for Bounded Univale Functions II. VI, 168 pages. 1982.

Vol. 914: M. L. Warshauer, The Witt Group of Degree k Maps ar Asymmetric Inner Product Spaces. IV, 269 pages. 1982.

Vol. 915: Categorical Aspects of Topology and Analysis. Procee ings, 1981. Edited by B. Banaschewski. XI, 385 pages. 1982.

Vol. 916: K.-U. Grusa, Zweidimensionale, interpolierende Lg-Spline und ihre Anwendungen. VIII, 238 Seiten. 1982.

Vol. 917: Brauer Groups in Ring Theory and Algebraic Geometry. Pr ceedings, 1981. Edited by F. van Oystaeyen and A. Verschoren. VI 300 pages. 1982.

Vol. 918: Z. Semadeni, Schauder Bases in Banach Spaces Continuous Functions. V, 136 pages. 1982.

Vol. 919: Séminaire Pierre Lelong – Henri Skoda (Analyse) Anné 1980/81 et Colloque de Wimereux, Mai 1981. Proceedings. Edi par P. Lelong et H. Skoda. VII, 383 pages. 1982.

Vol. 920: Séminaire de Probabilités XVI, 1980/81. Proceeding Edité par J. Azéma et M. Yor. V, 622 pages. 1982.

Vol. 921: Séminaire de Probabilités XVI, 1980/81. Supplément: Gé métrie Différentielle Stochastique. Proceedings. Edité par J. Azén et M. Yor. III, 285 pages. 1982.

Vol. 922: B. Dacorogna, Weak Continuity and Weak Lower Sen continuity of Non-Linear Functionals. V, 120 pages. 1982.

Vol. 923: Functional Analysis in Markov Processes. Proceeding 1981. Edited by M. Fukushima. V, 307 pages. 1982.

Vol. 924: Séminaire d'Algèbre Paul Dubreil et Marie-Paule Malliav Proceedings, 1981. Edité par M.-P. Malliavin. V, 461 pages. 19

Vol. 925: The Riemann Problem, Complete Integrability and Ari metic Applications. Proceedings, 1979-1980. Edited by D. Chudn sky and G. Chudnovsky. VI, 373 pages. 1982.

Vol. 926: Geometric Techniques in Gauge Theories. Proceedin 1981. Edited by R. Martini and E.M.de Jager. IX, 219 pages. 19